West Indian Green Monkeys: Problems in Historical Biogeography

Contributions to Primatology

Vol. 24

Series Editor:
F.S. Szalay, New York, N.Y.

Associate Editors:
P. Charles Dominique, Brunoy; *E. Delson*, Bronx, N.Y.;
W.L. Jungers, Stony Brook, N.Y.; *H. Kuhn*, Göttingen;
W.P. Luckett, Omaha, Nebr.; *J. Oates*, New York, N.Y.

Founding Editors:
A.H. Schultz †, Zürich; *D. Starck*, Frankfurt am Main

Basel · München · Paris · London · New York · New Delhi · Singapore · Tokyo · Sydney

West Indian Green Monkeys: Problems in Historical Biogeography

Woodrow W. Denham, Franconia, N.H.

6 figures and 8 tables, 1987

Basel · München · Paris · London · New York · New Delhi · Singapore · Tokyo · Sydney

Contributions to Primatology

Library of Congress Cataloging-in-Publication Data
  Denham, Woodrow W. West Indian green monkeys.
  (Contributions to primatology; vol. 24)
  Bibliography: p.
  1. Cercopithecus aethiops – Geographical distribution.   2. Cercopithecus aethiops –
  Migration.   3. Mammal populations – Barbados.   4. Mammal populations –
  Saint Kitts-Nevis.   5. Mammals – Barbados – Geographical distribution.   6. Mammals –
  Saint Kitts-Nevis – Geographical distribution.   7. Mammals – Barbados – Migration.
  8. Mammals – Saint Kitts-Nevis – Migration.   I. Title.   II. Series.
  QL737.P93D39 1987 [599.8′2 86-27675]

Contents

Illustrations

Figures

Tables

PREFACE

The primate literature contains several suggestions to the effect that studies of feral populations of African green monkeys on the West Indian islands of Barbados, St. Kitts and Nevis may yield useful information on adaptation and speciation in primate evolution. However, the history of these monkeys has been inadequately explored heretofore, and the explicitly and implicitly evolutionary work that has been conducted with them rests on shaky foundations.

This monograph is an exploratory study of primate historical biogeography, with equal emphasis being placed on history and biology. The work delineates but fails to answer questions concerning two major issues: migration of African monkeys to the West Indies (provenance, numbers, dates and routes of migration), and cyclic changes in the size of the Barbadian monkey population between 1627 and 1986. Most of the material that pertains to these issues appears in ambiguous historical sources that are not generally available to the scientific community, so relevant excerpts (i.e., the raw data) from those sources are included in a lengthy appendix.

The archival research that is reported here would have been impossible without the cooperation of the directors and librarians of the following organizations: in Barbados, the Ministry of Agriculture, the Department of Archives, the Barbados Museum and Historical Society, the Barbados Public Library, the Bellairs Research Institute of McGill University, the Caribbean Meteorological Institute, and the West Indies Collection of the University of the West Indies at Cave Hill; in London, the British Museum (Natural History) and the Public Record Office; in the United States, the Library of Congress, the Boston Public Library, and Baker Library at Dartmouth College; in Canada, libraries at the University of Toronto and McMaster University. I thank all of them for their enthusiastic support of the project.

Suzanne Skibiak translated the passage from Pere Labat's journal. Warren Alleyne allowed me to quote an "Act for Destroying Wilde Monkeys and Raccoons", dated 1679, that is in his possession. Thaya DuBois allowed me to include her unpublished demographic data in Chapter III. Jerome Handler, Michael McGuire and Peter Campbell directed me to valuable sources that I had missed in my own search of the literature, and allowed me to use unpublished information that they have accumulated in their own studies of social and natural history in the West Indies. My wife, Nancy Hubley, in interviews with Barbadians who migrated to Nova Scotia between 1910 and 1925, collected information for me concerning the distribution of monkeys in

Barbados early in the 20th century.   Jean Baulu and Julia Horrocks shared with me their extensive field experience with Barbadian monkeys, and Robert Speed told me a great deal that I
did not know about the geology of Barbados.   In Barbados, The
Bellairs Research Institute of McGill University and its
director Wayne Hunte provided an outstanding home-away-from-
home while the research was in progress.   I thank all of them
for their cooperation, generosity and assistance.

     I thank Peter Campbell, Christopher Hallpike, Jerome Hand-
ler, Michael McGuire, Richard Slobodin and Ronald Taylor, as
well as several anonymous reviewers, for constructively criti-
cizing an early draft of the monograph.   Peter Campbell, editor
of the Journal of the Barbados Museum and Historical Society,
was especially helpful to me as I prepared early versions of
Chapters II and III for publication as articles in his journal.
His firsthand knowledge of natural and social history of the
island, and of published and unpublished materials that are
available only on the island, saved me from important blunders.

     I thank my mother, Christine H. McElvaney, and the Univer-
sity System of New Hampshire for their contributions toward de-
fraying publication costs.

     Finally, I thank my wife for her enthusiastic participa-
tion in all aspects of the project.   Without her support, the
book would not have been published.

I. INTRODUCTION

Large feral populations of African green monkeys (Cerco-pithecus aethiops) live on the West Indian islands of Barbados, St. Kitts, and Nevis, and small numbers have been reported from Sint Eustatius. A population of African mona monkeys (Cer-copithecus mona) has been reported from the island of Grenada. Although all of these monkeys are cercopithecines, this book deals exclusively with the green monkeys.

St. Kitts, Nevis and Sint Eustatius are clustered together near the northern end of the Lesser Antilles. Barbados is lo-cated in the Atlantic Ocean about 600 km. to the southeast of that cluster, about 150 km east of Grenada and the Antillean Volcanic Arc, and about 4000 km west of the nearest point in Africa (Figure 1). Since there is no evidence that Old World monkeys lived in the Americas at the time of European discov-ery, it is reasonable to assume that African monkeys now living in the West Indies were introduced after European colonization began.

At least superficially, the situation in which the West Indian cercopithecines found themselves in earlier centuries was similar to that in which the ancestors of Darwin's finches found themselves upon arriving in the Galapagos Islands. Both were isolated on oceanic islands where they confronted few pre-dators, and could radiate into diverse and abundant unoccupied ecological niches. It was a situation that could, but need not necessarily, have given rise to rapid evolutionary change.

The potential importance of the West Indian populations as subjects of a "natural experiment" in primate evolution was re-cognized half a century ago, and a slow but steady stream of research has capitalized on the situation (Cameron 1930; Col-yer 1945, 1948; Ashton and Zuckerman 1950, 1951a, 1951b, 1951c; Ashton 1960; Sade and Hildrech 1965; Poirier 1972; McGuire 1974; Ervin and McGuire 1974; and several additional publications since the mid-1970's). In order to explore direc-tions and rates of change in morphology and behavior among various West Indian populations and between West Indian and African populations, people have needed to know where the mon-keys originated, when the first and last ones reached the West Indies, how isolated they have been since they arrived, and how the histories of the island populations differ from each other.

As early as 1948, Zuckerman requested detailed information about the monkeys' history from E. M. Shilstone at the Barbados Museum and Historical Society and C. E. Shephard at the Imper-ial College of Tropical Agriculture in Trinidad, but neither could assist him (letters dated August 1948 through February 1949 at the Barbados Museum and Historical Society). Readily available but superficial historical sources suggested that the

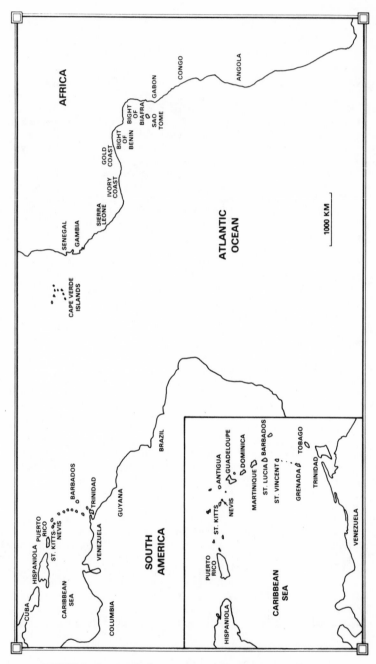

FIGURE 1.   Africa and the West Indies.

provenance, antiquity, and isolation of the West Indian populations were not problematic, and a simple historical account
gradually emerged in the scientific literature.

But in their eagerness to deal with scientific, biological, ultimately evolutionary problems, early researchers placed
insufficient emphasis on the fact that history on a "human time
scale" of decades and centuries - not on a geological time
scale or a laboratory time scale - constitutes both the context
and the raw material of evolution in action. Failure to fully
grasp this point introduces large elements of uncertainty and
guesswork into any evolutionary study. Such is the case with
the West Indian green monkeys, where the obvious plausibility
of the currently accepted historical account has come to obscure questions about its validity.

Objectives

My objectives in the research reported here are both substantive and methodological. My substantive objectives are to
introduce documentary and observational data that shed new
light on two facets of the history of West Indian green monkeys, and to suggest several plausible interpretations of their
history that should be evaluated in conjunction with future
evolutionary studies of the animals. Most of the historical
scenarios that I suggest must be wrong, for I have attempted to
present them as mutually exclusive competing hypotheses insofar
as that has been possible. However, none of them can be ruled
out until they have been tested rigorously in the future.

My methodological objective is to demonstrate some of the
many historiographic problems that must be confronted and re-
solved before the history of the West Indian green monkeys, or
any other introduced primate population, can be proffered with
any degree of confidence. The history of research with West
Indian green monkeys reveals a considerable number of scientists who have had their critical faculties finely tuned with
regard to the quality and limitations of their scientific data,
but at the same time have been entirely uncritical of historical materials that have been essential to their work, and that
have been even more difficult to obtain and interpret than
their conventional biological and behavioral data. Here I attempt to demonstrate that problems in primate historical biogeography can be approached from various directions, and are
neither trivial nor insurmountable.

An anonymous reviewer aptly described the monograph as a
"historical whodunit on a biological subject". I could not have
said it better.

Data

In attempting to reconstruct the history of green monkeys
in the West Indies, I have relied primarily on documents from
the 17th through 20th centuries, and have supplemented them
with data from field observations and interviews that were conducted in 1977 and 1978.

Observational and interview data pertain exclusively to
Barbados. Although scanty, they are numerical or quantitative
and may contribute to more sophisticated studies of Barbadian
monkey populations.

The documentary sources are of many kinds. They include
the following: 1) reports by natural historians, physicians,
biologists, anthropologists, and other trained scientists who
commented on monkeys in the West Indies from the 17th through
20th centuries; 2) observations of monkeys in the West Indies
by sailors, travelers, historians and other nonscientists; 3)
scientific papers from the 18th through 20th centuries that
pertain to the distribution of green monkeys in West Africa and
the Cape Verde Islands; 4) recent histories of the West Indies
and the slave trade; 5) Acts of the Legislature of Barbados;
6) reports by government agencies and Barbadian newspapers concerning monkeys in the West Indies; and 7) European social and
intellectual histories that deal with monkeys as menagerie exhibits and pets in Europe and the West Indies.

My search for documentary information has focused almost
exclusively on published sources, and I have not attempted to
explore the enormous volume of unpublished material that may
contain relevant and important information. Typically the published references are vague and ambiguous, but they contain the
evidence upon which I have based my inferences and interpretations. Since the materials are difficult to interpret, since
many of them are rare and inaccessible, and since their rein-
terpretation may be necessary when unpublished sources have
been examined, I have quoted the sources extensively in an appendix. Hence the data are available to display by themselves
their ambiguity, opacity, and possible inadequacy for the tasks
at hand. In several instances, the sources are just as important for what they omit as for what they include, and I have
quoted in full or at considerable length in order to indicate
precisely what is not available.

Quotations in the appendix appear in alphabetical order by
author's last name. Citations of items that are quoted in the
appendix appear as (SCHOMBURGK 1848), while citations of items
that do not appear in the appendix appear as (Ligon 1657).

In addition to their potential value for further work on
the problems that I address here, the items in the appendix
contain a wealth of information that I have not explored fully
in the text. Their availability here should facilitate research into other aspects of primate historical biogeography in
the West Indies.

Many unpublished documents are likely to repay the efforts of anyone who undertakes the time-consuming task of examining them for information about monkeys. They include wills, deeds, plantation records and government documents preserved in the Department of Archives in Barbados and the Public Record Office in London; ships' logs and letters in libraries and museums in Barbados, England, the Maritime Provinces of Canada, and the New England region of the United States; the Edgerton and Sloan manuscripts at the British Museum; and a volume of the Lucas manuscripts, purportedly dealing with the Barbadian monkeys in the 17th and 18th centuries, that is missing from the Barbados Public Library.

In order to make sense of early reports that pertain to West Indian monkeys, it is necessary to confront two conceptual problems in primate historiography. The first is the importance of monkeys in European and African folklore, and the second is the history of changes in European views of monkeys since the beginning of the 17th century. Failure to consider both of them adequately can yield extreme disorientation with regard to scientific and folk taxonomies, and serious limitations on one's ability to distinguish among useful observations, local anecdotes, African folklore, European folklore, and European social philosophy.

McDermott (1938) summarizes an enormous body of literary, artistic, philosophical and scientific material that pertains to monkeys in Egypt, Greece, the Roman Empire, and adjacent regions prior to the fall of Rome. Janson (1952) extends McDermott's artistic and literary inquiry into Medieval and Renaissance Europe, and Hodgen (1964:386-432) examines the place of monkeys in European social philosophy in the 16th to 18th centuries. Additional materials on the history of European views of monkeys are contained in standard works in systematic zoology such as Linnaeus (1766) and Buffon (1767). To my knowledge, no comprehensive study of the monkey in African and Afro-American folklore has been published, but voluminous West Indian raw materials, especially Annancy (or Anansi) stories in which monkeys are frequent characters, are readily available in Beckwith (1924), Jekyll (1906), Parsons (1925, 1933, 1936, 1943) and many others, and studies of African folklore are legion. Although I mention these publications only rarely or not at all in the pages that follow, they are essential background reading for anyone who wishes to pursue the subject further.

Chapter II deals comprehensively with migration of green monkeys from Africa to Barbados, St. Kitts and Nevis, for the materials that I have found are appropriate to that broad purpose. Chapter III deals exclusively with changes in the Barbadian monkey population during the last 350 years and does not pay equal attention to that matter on the other islands, for none of the sources that I have found suggest that significant changes have occurred in the non-Barbadian populations. Chapter IV compares Barbados with St. Kitts, and comments briefly on the value of doing "human scale" history in primatology.

There are many unexplored questions whose answers could
have made this a more complete first contribution toward a historical biogeography of the West Indian cercopithecines, but
investigating them was not possible.   Hence, the monograph is
offered as a stimulus and guide for further research, not as a
definitive history of these animals.   Most of the answers   -
probably even most of the questions - remain to be found.

II. CERCOPITHECINE MIGRATION TO THE WEST INDIES

The Problem

There are minor differences of opinion in the literature, but it has been commonly accepted that Cercopithecus aethiops sabaeus is the only nonhuman primate species that has successfully colonized Barbados, St. Kitts and Nevis, that those monkeys reached the West Indies in the mid-17th century, that they originated in Senegal or Gambia, that only a few of them survived the difficult "middle passage" aboard French slave ships, and that the island populations have remained genetically isolated from African populations and from each other since the beginning of the 18th century. This migration scenario appears, in bits and pieces, in the following quotations:

> The Green Monkey (Cercopithecus aethiops sabaeus), whose normal habitat is West Africa, was introduced into the West Indian island of St. Kitts by the French when it was their possession... It has been locally established for quite 250 years (Colyer 1945:845).

> Some 300 years ago, free-ranging colonies of green monkeys of the West African species Cercopithecus aethiops sabaeus became established on the Caribbean island of St. Kitts (Ashton 1960:563).

> African Green Monkeys (Cercopithecus aethiops sabaeus) were introduced into St. Kitts, West Indies in the late 17th century as a by product of the slave trade between Senegal and the West Indies (Sade and Hildrech 1965:67).

> Green monkeys (Cercopithecus aethiops sabaeus) have inhabited the West Indian island of St. Kitts for approximately 300 years (Poirier 1972:20).

> This Old World monkey (was) brought to the Caribbean in the mid-seventeenth century by early French settlers from Senegambia (Ervin and McGuire 1974:36).

> Over 300 years ago the vervet was brought from West Africa to (St. Kitts) ... where it has since remained self-perpetuated; this original stock (of 50 to 250 animals) is essentially free from the infusion of additional monkeys (McGuire 1974:5-7).

The contexts from which these quotations were extracted indicate that the assertions are tentative rather than defini-

tive, but as the same tentative assertions have been repeated
in slightly altered forms through the years they have gradually
achieved the status of self-evident truths that merit no further attention.

However, the authors just quoted, as well as others, have
published field and laboratory observations that cast at least
a shadow of a doubt on the validity of the historical account
presented above.   One conspicuous problem is that some of the
observations raise questions about the identity of the Kittitian monkeys.

Sade and Hildrech (1965), Poirier (1972) and McGuire
(1974), in discussions of the taxonomic status of Kittitian
monkeys, have classified the animals as Cercopithecus aethiops
or Cercopithecus aethiops sabaeus.   Outside the West Indies,
C. aethiops has been reported "in savannah regions throughout
(sub-Saharan) Africa, from Senegal to Ethiopia, and from the
Sudan to the tip of Africa" (Tappan 1960), and in the Cape
Verde Islands (Napier and Napier 1967).   In conjunction with
its broad geographical distribution, there are significant
regional variations in its appearance.

Some people argue that C. aethiops is a superspecies that
subsumes several species and subspecies, while others disagree.
Pocock (1907), Schwarz (1936), Booth (1956b) and Napier and
Napier (1967) propose schemes similar to that of Dandelot
(1959), who suggests that the superspecies C. aethiops be sub-
divided as follows:

Genus Cercopithecus
 Superspecies C. aethiops
  Species C. a. aethiops, Grivets: River Volta to Ethiopia
    Subspecies C. a. aethiops aethiops
    Subspecies C. a. aethiops tantalus
  Species C. a. sabaeus, Green Monkeys: west of River Volta
  Species C. a. pygerythrus, Vervets: southern third of Africa
    Subspecies C. a. pygerythrus pygerythrus
    Subspecies C. a. pygerythrus cynosurus

On the other hand, Struhsaker (1970:402) agrues that C.
aethiops is "one species polymorphic for pelage color and pattern", and Tappan (1960) says it is "a single interbreeding
species with only temporarily isolated local populations".

Despite the taxonomic problems associated with these animals, there is no disagreement concerning either the diversity
of appearance among C. aethiops in various parts of Africa or
their ability to interbreed freely, nor is there any doubt that
the typical appearance of the West Indian green monkeys is
closer to that of Dandelot's C. a. sabaeus than it is to other
members of the species or superspecies.

None of the evidence that I have found suggests unequivocally that the Kittitian monkeys have ever been identified as
members of any genus other than Cercopithecus, but the animals
present problems of identification at both the species and

subspecies levels.   SCLATER (1866), on the basis   of   three
specimens delivered to the Royal Society in London,   identified
them as Cercopithecus callitricus,   Geoff., (i.e.,   C.   a.
sabaeus),   and ELLIOT (1905) and ALLEN (1911) concurred.
However, HOLLISTER (1912) reported that the skin of an African
mona monkey (C. mona) was procured in St. Kitts in 1880 and de-
posited at the U. S. National Museum along with specimens of
green monkeys from the same island.   Hall and Kelson (1959) repeated Hollister's report and described skins of C. a. sabaeus
and C. mona from St. Kitts.
     Since feral populations of mona monkeys have been reported
from the West Indian island of Grenada (Allen 1911),   perhaps
the mona skin came from Grenada or elsewhere and was mistakenly
attributed to St. Kitts.   Indeed, F. A. Ober, the man who procured the specimen that Hollister reported,   devoted an entire
chapter in one of his many West Indian guidebooks to the hunting of an unidentified species of monkey in Grenada (Ober 1880:
263-279).   But mistaken provenance is not the only plausible
explanation.
     Poirier (1972:21-22) reports color variations among dif-
ferent monkey populations on St. Kitts, and between populations
on St. Kitts and Nevis.   Typically, the St. Kitts monkey is
greenish-gold on the dorsal surface and yellowish-white on the
ventral.   Its black face is bordered by white hair, and the
color of its tail ranges from greenish-gold at the proximal end
to yellow or reddish-yellow at the distal end.   Poirier suggests that most of the observed devistions from this pattern
probably exemplify a range of intraspecific color variability.
But he adds:

     "Some   knowledgable Kittitians report an   anomalous
     population residing in one of the mountain ravines.
     Several specimens from the high mountains exhibited
     a mottled facial coloring,   particularly about the
     eyes,   and a thicker, somewhat darker, hair coat...
     The   speckled facial characteristics of the one
     mountain population are ... quite different (from
     those of the typical St. Kitts monkey)".

Sade and Hildrech (1965:70-71) note the following in their
discussion of Hollister's report:

     "We were told that monkeys seen high in the   moun-
     tains (of St. Kitts) were whiter than monkeys com-
     mon at lower elevations. At about 2000 feet eleva-
     tion ... Sade saw a group which included monkeys
     with browner backs and monkeys with whiter backs.
     In other pelage characters, and in form of face
     they were all of aethiops type. ... (It) would be
     premature to state that C. mona is not established
     on St. Kitts, (but) C. aethiops is certainly by far
     the most abundant."

Conversely, McGuire (pc) on the basis of several years of research in St. Kitts, says that the differences reported by Poirier, Sade and Hildrech between the monkeys of St. Kitts and Nevis, and between those at low and high elevations on St. Kitts, disappear when the animals are put in the same environment for six months. But McGuire also notes that the mona skin attributed to St. Kitts has now been reclassified as aethiops, which suggests that the range of variation in pelage characters within the Kittitian population is truly enormous, or that the original classification of the skin was rather inept.

Parasitological and morphological studies of St. Kitts monkeys are summarized clearly by Poirier (1972:21):

> Cameron (1930) reported that a large sample of intestinal nematode parasites collected from the St. Kitts monkeys revealed several minor differences from those collected from African hosts. Cameron concluded that a different species of parasite infected the St. Kitts monkeys ... (but cautioned) that this new species may have died out in Africa or simply may not have been represented in that sample. ... Colyer (1948) reported a high incidence of dental anomalies among ... 95 complete skulls of St. Kitts monkeys when compared with a similar African sample. Ashton and Zuckerman (1950, 1951a, 1951b, 1951c), using Colyer's specimens (compared) St. Kitts green monkeys (with) African cercopithecus sp. (T)he skulls and teeth of the St. Kitts monkeys were larger and less variable, although within the normal range of other cercopithecidae. Ashton (1960) concluded that while variations existed, they were insufficient for taxonomic division higher than the subspecific level.

But in a potentially important statement that Poirier omitted, Ashton (1960:373) noted that, if the apparent rate of change in tooth size continued unabated in the St. Kitts monkey, "its teeth would double in size in approximately 5,000 years."

Taxonomic affiliations of Barbadian green monkeys seem to be somewhat less problematic than those of the Kittitian monkeys. However, Horrocks (1984:33) argues that a population of C. a. tantalus may have existed on the island until recently, since the monkeys in Barbados, and some of those in St. Kitts (McGuire 1974) have a faint narrow white browband that is more typical of tanatlus than sabaeus.

The anomalous findings introduced here may be interpreted in at least four fundamentally different ways: a) as evidence of extreme polymorphism within a single species; b) as evidence of subspeciation that has begun as a result of a founder effect combined with three centuries of genetic drift or gradualistic natural selection; c) as evidence of a complex history of migration that has yielded monkeys of mixed parentage; or d) as

a   result of erroneous observations,   poor sampling techniques,
and defective data analysis.

Interpretation a) assumes that the apparent anomalies   are
not anomalies at all, but are simply manifestations of a normal
range of phenotypic variation within a single species.   Interpretation b) assumes that they are real anomalies caused by
microevolutionary processes leading to differentiation of isolated populations from each other and from their parent   stock.
Interpretation   c) assumes that the anomalies are due to interbreeding   and   hybridization   among   animals from   two   or   more
parent populations.   Interpretation d) is intrinsically noncom-
mital about the monkeys.

Interpretations a), b), and d) have been advanced by those
who reported the observations.   In the remainder of this chapter,   I examine the   plausibility of alternative c).   My inten-
tion   is not to eliminate the other three   from   consideration,
nor to insist that   the interpretations are mutually exclusive,
but   rather to give careful   consideration to a group of interpretations that has been ignored heretofore.

Since cercopithecine migration to the West Indies is inti-
mately connected with the political, economic, military and social history of European colonialism in the Caribbean,   I shall
review   a broad range of materials that typically   lie   outside
the   domain of primatology.   They include the early history of
English and French settlement of St. Kitts, Nevis and Barbados;
African   origins of West Indian   slaves;   transatlantic   trade
routes in the 17th through 19th centuries;   the role of Barbados   as a slave entrepot;   monkeys as pets aboard ships and as
trade   goods   sold to menageries in Europe and the   New   World;
bounty laws and import restrictions against monkeys in the West
Indies;   and early observations of monkeys in the West   Indies
and the Cape Verde Islands.

Provenance of Migrants

St.   Kitts was discovered by Columbus in 1493, but was not
settled   by   Europeans   until   1624 when   it   became   the   first
British colony in the West Indies.   In 1627,   French colonists
moved onto each end of the island,   while the British   retained
control of the middle of it.   That arrangement persisted until
1665 when war erupted between England and France.   In the next
half-century, control of the island changed from one country to
the   other   seven times until the British finally   acquired   it
permanently in 1713 (Dunn 1973:117-149).

In comparison with St.   Kitts, Barbados had a stable colonial history.   The Portuguese were the first Europeans to dis-
cover it,   in 1536, but they did not settle there.   The British
then   claimed   it in 1625 and settled it as their second   West
Indian   colony in 1627.   It remained a British colony   without
interruption until it achieved independence in 1966.

The most frequently cited reference concerning the trans-atlantic migration of cercopithecines appears in a journal kept by Pere Jean-Baptiste Labat during a visit to St. Kitts in 1700, when the island was under French control. According to LABAT (1931:42), ships reached the West Indies from West Africa carrying, among other goods, slaves and monkeys. He did not specify the locations from which the ships left Africa, but the co-occurrence of C. a. sabaeus and 17th century French col-onists in Senegal and Gambia (sometimes called Senegambia) led Sade and Hildrech (1965:67) and Ervin and McGuire (1974:36) to identify those countries as the most likely homes of the origi-nal West Indian green monkeys.

These inferences are strongly supported by "the Sieur D. B." (OLIVER 1897:9), a natural historian who sailed past Sene-gal enroute to Madagascar from Europe in 1669. He noted that French settlements in Senegal belonged to the gentlemen of the Company of the West Indies who traded in the same goods that Labat listed, including monkeys. Unfortunately, Labat and the Sieur D. B. did not identify the species of monkeys that the French West Indian Company exported from Africa, they said nothing about monkeys that may have been exported to the West Indies from other parts of Africa, and their remarks on French trade probably do not pertain to Barbados.

The only report that I have found that identifies the pro-venance of West Indian cercopithecines independently of Labat is MORRIS (1914:376), who provides no source for his assertion that the monkeys of Barbados and St. Kitts were introduced from Sierra Leone during the period of slavery. Sierra Leone was a Portuguese colony from 1492 until the British took possession of it at the end of the 17th century, and is adjacent to Sene-gal and Gambia. Morris seems to have escaped the attention of primatologists, for his assertion has not been repeated in the primatological literature.

Since the information that is available from people who could have seen where the monkeys originated and where they went is inadequate, we must attack the problem indirectly. If we accept the statements by Labat and the Sieur D.B. to the ef-fect that monkeys were transported by slave ships, one indirect approach to the question of provenance it to investigate the 16th to 19th century Atlantic slave trade.

Where did West Indian slaves come from? This is an enor-mous and extraordinarily complex question that Sade and Hild-rech (1965:71) glossed over when they said, with no further discussion of the issue: "It was from Senegal that slaves were brought to the West Indies." Although that statement agrees with Labat's report and is compatible with the known distribu-tion of C. a. sabaeus, we must pursue the matter further.

Very little is known about the French West Indian slave trade in general, and partly because of the chaotic political history of St. Kitts during the 17th century, the origins of its slaves in particular are virtually unknown. Even by infer-ring from the comments of Labat and the Sieur D. B. that Sene-

gal was a major supplier to French settlers on St. Kitts in the
17th century, we still do not know the volume of the French
trade or the identities of other locations in Africa that supplied slaves to French settlers in the New World. Curtin
(1969:121) notes that few French slaves were transported in
French ships.

On the other hand, a great deal is known about the British
slave trade in general and about the Barbadian trade in particular. Barbados was the final destination of approximately
360,000 slaves who were shipped from Africa between 1627 and
1834. Approximately 20,000 to 30,000 arrived by 1650, and from
2,000 to 4,000 arrived each year thereafter until 1774. From
1774 to 1807, fewer than 400 arrived each year, and no more
than fifty arrived after 1807 when the British slave trade of-
ficially ended (Handler and Lange 1978:22).

Until the 1660's the Dutch transported most of the slaves
that went to Barbados. Virtually none of those slaves originated in Senegal or Gambia, two percent originated on the Gold
Coast, 64 percent on the Bight of Benin, and 34 percent on the
coasts of Gabon, Congo and Angola. From 1680 until 1807, British slavers supplied their colonies in the Caribbean with
people from the Windward and Gold Coasts and the Bights of
Benin and Biafra. During that period, less than ten percent of
the British slaves originated in Senegal and Gambia (Handler
and Lange 1978:24-25; Curtin 1969).

In addition to the 360,000 slaves whose final destination
was Barbados, thousands more stopped at the island only
temporarily. Because of its location as the easternmost of the
West Indian islands and its early reputation as the wealthiest
and most powerful of the British sugar colonies, Barbados was
the first stop in the New World for many slavers and other
ships enroute from Africa to the West Indies and North America,
and from England via Madeira or the Cape Verde Islands to the
West Indies and Guyana. It was the premier British slave
entrepot in the Caribbean from the mid-17th century until the
final suppression of the slave trade (Coke 1810:127-128; Dunn
1973:234-238; Sheridan 1973:125; for references to British
Parliamentary Papers that report numbers of slaves imported and
re-exported from Barbados between 1789 and 1830, see Handler
1971:107-109). No one knows how many slave ship voyages either
ended at Barbados or stopped there enroute to other ports, including St. Kitts, but it is likely that there were many thousands of them. If monkeys reached Barbados on slave ships,
they had innumerable opportunities to make the voyage, and they
could have come from anywhere on the coast of Africa south of
the Sahara.

Further examination of British slave data, in conjunction
with maps of the distribution of C. aethiops in Africa (McGuire
1974; Dandelot 1959; Booth 1956a, 1956b), reveals that both
sabaeus and tantalus are present on the Gold Coast where a
large percentage of British slaves originated. Since members

of both subspecies were commonly exported to England (POCOCK 1907:727, 731), it is reasonable to suspect that representatives of both went to the West Indies as well. Furthermore, the subspecies pygerythrus and cynosurus are present in Angola and Mozambique. Since as many slaves reached Barbados from those two areas as from Senegambia (Handler and Lange 1978; Curtin 1969), it is possible that representatives of the southern subspecies of C. aethiops reached the West Indies where they interbred with representatives of the northern subspecies.

In general, the slave trade data are compatible with the notion that green monkeys reached St. Kitts from Senegal and Gambia when part or all of the island was under French control during the 17th century, but they are by no means unequivocal on that point. They definitely do not suggest the same provenance for the Barbadian monkeys. Since a great majority of Barbadian slaves originated outside of Senegal and Gambia, the data suggest that the Gold Coast and points east and south are more likely sources of the Barbadian monkeys, and that those monkeys may have interbred with members of other C. aethiops subspecies from elsewhere in Africa. Interbreeding of comparable nature and magnitude may be responsible for the anomolous appearance of some of the Kittitian animals, and for the reported morphological differences between Kittitian and African populations.

Although the evidence introduced here suggest that the West Indian monkeys had heterogeneous origins, there is yet another possibility. The range of C. a. sabaeus includes the Cape Verde Islands, where presumably they were introduced from Africa no later that the mid-17th century (Napier and Napier 1967:100), and perhaps much earlier. In 1673, during a visit to St. Jago Island in the Cape Verdes, FRYER (1909:40) met natives on the beach selling monkeys, "such green ones as are commonly seen in England to be sold." George Edwards, whose fanciful illustration appears on the cover of this monograph, said that seamen of his day generally referred to green monkeys as St. Jago monkeys because the animals were brought to England from St. Jago in the Cape Verdes (EDWARDS 1758:10). Linnaeus' (1766) type specimen for C. sabaeus came from St. Jago, and during the first half of the 19th century, both JARDINE (1833:203) and GOLDSMITH (1840:501) noted that green monkeys lived in the Cape Verde Islands.

The presence of C. a. sabaeus in the Cape Verdes raises the possibility that the monkeys used those islands as stepping stones to the West Indies. The Cape Verdes were a vital link in the British "triangle trade" (manufactured goods to Africa, slaves to the West Indies, sugar to England), and the Cape Verdians shipped salt and livestock directly to Barbados aboard English ships that did not go to Africa as part of the triangle trade (Duncan 1972; PRO Doc. No. T.64/48). Perhaps green monkeys were introduced to the Cape Verdes between the 1460's when Portuguese settlement began there, and the 1620's when British settlement began in Barbados, St. Kitts and Nevis. If the Cape

Verdes then served as the source of monkeys that became established in the West Indies, it is possible that the West Indian
populations arose from a much more narrowly circumscribed
founder population than has been suspected heretofore.

On the basis of slave trade statistics and the wide distribution of C. aethiops in Africa, heterogeneous origins are
most likely, but on the grounds of geographical proximity, a
Cape Verdian origin of the West Indian monkeys cannot as yet be
ruled out.

Number of Migrants

The next issue to be considered is the number of monkeys
that survived the passage from Africa to the West Indies.
McGuire (1974:7) speculated that 50 to 250 may have arrived
safely in St. Kitts from Africa, but I suspect that his estimate is too conservative for St. Kitts, and much too small for
the West Indies as a whole.

There were at least two important reasons for monkeys to
sail from Africa to the West Indies. Some went as trade goods,
some as sailors' pets, some as both. Although it is conceiv-
able that others were used a food for slaves in transit, I have
found no evidence to that effect. Information concerning numbers of monkeys exported from Africa falls into three classes:
general comments on monkeys in Europe, reports of shipments of
mixed or unknown species, and specific comments on the importance of green monkeys as pets. Since most of the evidence is
circumstantial, I discuss it as some length. Its quantity is
more informative that is its quality.

McDermott (1938) and Jansen (1952) leave no doubt about
the importance of monkeys in ancient Egypt and Mesopotamia,
Greece and Rome, and Medieval and Renaissance Europe. Among
the monkey species that were discussed, disected, traded, wor-
shipped, and artistically depicted in the civilizations of antiquity were baboons (both Papio cynocephalus and P. hama-
dryas), Barbary apes (Macaca sylvanus) and rhesus macaques (M.
rhesus), Hanuman langurs (Presybtis entellus), and cercopithecines (C. aethiops). Baboons were sacred in Egypt for about
4000 years, and the others were common as pets and entertainers. Of particular interest here are two very early, detailed,
naturalistsic scenes from the House of the Frescoes at Knossos
(ca. 1550 BC) and a tomb at Thebes (ca. 1450 BC) that unambiguously depict aethiops and tantalus (McDermott 1938:4-5, 23-24,
274-275).

Between the fall of Rome and the High Middle Ages, monkeys
were rare in Europe. Beginning in the 12th century, however,
Barbary apes and baboons appeared in Spain, Italy, France, Germany and England as entertainers with minstrels and jesters, as
expensive pets of the nobility, and as religious, philosophical
and social symbols in art and literature (Jansen 1952; Burke
and Caldwell 1968). The "enthusiasm for the collection of

curios" (Hodgen 1964:114) that characterized the Renaissance gave rise to museums, botanical gardens, and menageries all over Europe (Hodgen 1964:117; Burckhardt 1960:217; Jansen 1952:149) that created a demand for monkeys as exhibits. As a result of Renaissance voyages of exploration, longtailed monkeys from West Africa reached Europe in the 15th century, and New World monkeys arrived in the 16th. When New World capucins became favorites of organ grinders (Hall and Kelson 1959:223), they replaced the less tractable Barbary apes that earlier generations of minstrels and jesters had owned. By the time Barbados and St. Kitts were settled in the 17th century, captive monkeys had been commomplace in Europe for five hundred years.

I have found no social or maritime histories that deal comprehensively with monkeys as sailors' pets, but sailors sometimes discussed their pet monkeys in their journals. For example, Hugh Crow, the captain of the last slave ship to leave Liverpool before the British slave trade ended in 1807, described hilarious problems that he once had with a large monkey of unreported species that he bought at Bonny on the Bight of Biafra and subsequently sold in Kingston, Jamaica (WILLIAMS 1897:647-648). On another voyage from Bonny, Crow's ship stopped at the Portuguese island of Sao Tome in the Gulf of Guinea and later departed with "additional passengers in the shape of several monkeys presented by the Governor to Captain Crow" (WILLIAMS 1897:651). One of the Sao Tome monkeys, which were of "different sizes and species", sickened and died before reaching Jamaica, but it appears that the others survived (WILLIAMS 1897:654).

I have located no statistical data concerning 17th to 19th century trade in monkeys, but the animals' common occurrence as pets and as displays in public and private menageries suggests that they may have been shipped in large quantities for sale long before they became important for medical research and drug production. LABAT (1931:42-43) recounted a story (perhaps apocryphal) about a ship that sailed from Africa to France in the 17th century carrying 330 monkeys in cages and boxes. The captain thought it was peculiar to receive such a large order, and when he reached France he discovered that he should not have transported so many. But by then he had complied with the order and delivered more than 300 monkeys on a single trip. WILLIAMS (1897:384) reports that a British ship enroute to Liverpool from Demarara (Guyana) in 1801 was attacked by a French privateer that plundered its cargo which included, among other goods, a large collection of birds and monkeys. It is not clear from the report where the attack occurred, whether the monkeys were Old or New World species (though the departure from Demarara suggests that they were New World monkeys), or what happened to them. Finally, on 20 October 1905, a Bridgetown, Barbados, newspaper (Anon. 1905) reported that 200 monkeys had recently escaped from their cages aboard a ship enroute from Japan to New York, but no details were given.

Relative frequencies with which various monkey species were acquired as pets and exhibits are not known in detail, but it is clear that green monkeys from St. Jago were especially popular. FRYER (1909:40) said they were commonly seen in England in the 17th century; EDWARDS (1758:10-11) said that they were commonly brought into most of the maritime trading ports of Europe, and described one that was kept as a pet by the Duke of Richmond in Whitehall, England; JARDINE (1833:140) said that the green monkey was one of the most abundant cercopithecines and was perhaps oftener seen in captivity than any of the others; and GOLDSMITH (1840:407) said it was one of the species most frequently imported into Europe. POCOCK (1907:727) agreed that the green monkey was one of the commonest in European menageries, but also noted that a large number of specimens of C. a. tantalus had been exhibited in the Gardens of the Zoological Society of London (POCOCK 1907:731).

The fragmentary evidence that is assembled here creates a general impression of large-scale, long-term migration of monkeys, but yields no firm estimate of the number of monkeys that reached the West Indies from Africa or South America. The animals could have arrived singly or in small groups, and they could have arrived rarely or frequently. Nevertheless, the demand for monkeys in general, and for green monkeys (and presumably capucins) in particular, combined with the vast number of ships that stopped at Barbados and St. Kitts enroute to Europe from Africa and South America, suggest to me that McGuire's guess of 50-250 animals is much too conservative. Even if the mean migration rate was as low as one monkey per twenty-five voyages - a figure for which I have no supporting evidence - thousands of monkeys could have reached the West Indies after 1624.

Dates of Migration

According to the currently accepted historical account, green monkeys became established on Barbados, St. Kitts, and Nevis soon after 1650 and their migration from Africa ended soon after 1700. These dates are based on several disparate bits of information. a) Richard Ligon, a historian who lived in Barbados between 1647 and 1650, did not mention monkeys in his chapter on the natural history of the island (Ligon 1657). b) LABAT (1931:183}, during his visit to St. Kitts in 1700, said that pet monkeys escaped from French houses during the French-English conflict on the island. Presumably that happened near the beginning of hostilities in 1665. c) The Legislature of Barbados stated that monkeys were pests and passed a bounty law against them in 1682 (PRO-CSPCS 1682). It has been assumed that the bounty law was accompanied or quickly followed by a law against further importation of monkeys into Barbados. Furthermore, Labat said that monkeys were pests in St. Kitts in

1700, and it has been assumed that Kittitians responded to them in the same way that the Barbadians did. Although the sources cited here have been reported correctly elsewhere, other documents raise questions about the inferences that have been drawn from them.

When did the migration begin?

First, we must consider Ligon's (1657) failure to mention monkeys in his chapter on the natural history of Barbados. The importance of that fact as evidence for the absence of monkeys from Barbados seems to be enhanced when we note that Captain John Smith, who visited Barbados shortly after the first settlement was established there in 1627, published a journal of his travels (Smith 1627) that also omits monkeys from a list of Barbadian animals and that the same omission occurs in Clarke's (1670) lengthy descriptions of Barbadian natural history.

But we also know that monkeys were so troublesome in Barbados by 1680 that a series of bounty laws was passed to reduce their numbers (see below), yet several works that were published after 1680 and that might reasonably be expected to mention monkeys fail to do so. They include books by Labat (1722, 1931) who visited Barbados in 1700 during the trip that eventually took him to St. Kitts where he talked about monkeys at some length, Sir Hans Sloane (1707) who visited Barbados and contributed major works on the natural history of the West Indies, and George Pinckhard (1806) and John Poyer (1808) who wrote important histories of Barbados. The fact that none of these post-1680 writers mentioned monkeys when we know that monkeys were or had been important pests seriously weakens the argument that Ligon's omission means that monkeys were absent from the island when he worked there.

The argument based on the absence of evidence in Ligon's book is at least challenged if not directly weakened by Schomburgk's (1848) classic history of Barbados which states unequivocally that the first settlers found monkeys in large numbers upon their arrival. Since Schomburgk knew his predecessors' works, his assertion explicitly rejects their omissions of monkeys from their animal lists. Unfortunately, Schomburgk did not cite any sources in support of his position, and I have been unable to find any pre-1848 works that directly corroborate it.

However, an early item of circumstantial evidence that was brought to my attention by P. Campbell (pc) seems to offer some support for Schomburgk's position. It is a deed (BDA.RB3/1:148), dated 18 July 1642, that relates to the sale of eighteen acres of land at Apes Hill in St. James Parish, about 4 km. northeast of Holetown, the site of the first British settlement in Barbados. Since "ape" was the common term for monkeys in the 17th century (Jansen, passim), and since Apes Hill is in an area that is preferred by monkeys now (see Chapter III), its name may mean that monkeys lived there by 1642. Ligon's failure to mention the monkeys, if they existed, per-

haps was due to his failure to visit many parts of the island
(Watts 1966:41) and his corresponding ignorance of the monkeys'
existence.

The earliest unambiguous reference that I have found
concerning monkeys in Barbados is a letter written in 1671 in
which the author said that he was sending a monkey from
Barbados to a friend in England (PRO-CSPCS 1671). Since he did
not say whether the monkey originated in Barbados or elsewhere,
the letter provides no evidence concerning feral monkeys in
Barbados in 1671.

The 1682 bounty law that is cited frequently in discussions of West Indian cercopithecines is only one of several
that the Legislature of Barbados passed over a period of almost
forty years. Rawlin (1699:120) noted that a Barbadian bounty
on monkeys expired in 1678. Since some of the later bounty
laws were in effect for two years each, that one may have been
passed as early as 1676. However, since much of Rawlin's work
has been challenged by Hall (1764) and others (P. Campbell,
pc), it is possible that his date is incorrect.

The first bounty law of which there is an unequivocal record was introduced in the House of Assembly in December 1679.
The House's decision that it should lie under consideration until the next sitting suggests that there had been no previous
bounty law. An Act for Destroying Wild Monkeys and Raccoons,
which provided a five-shilling bounty for each monkey's head
delivered to a parish churchwarden, was passed by the House
early in 1680 and became law in March 1680. It was continued
in force by an act of July 1682 and again by an act of November
1684, at which time its previous two year limitation was re-
moved and it became permanent. In March 1714 the act was revised to provide a ten-shilling fine against any churchwarden
whenever he refused to pay the five-shilling bounty out of tax
funds (PRO-CSPCS 16 December 1679, 17 February 1780, 1 July
1682, 6 November 1684, 4 March 1714; HALL 1764:106, 111-112,
215). Clearly monkeys were perceived as pests in the 1670's
and thereafter, and that fact argues for their introduction
prior to 1670.

Labat's journal, which suggests that the Kittitian monkeys
escaped from their owners in 1665, is the only primary source
that I have found concerning Kittitian monkeys in the 17th century. The next reference is by William Smith, a resident of
Nevis, who visited St. Kitts in 1719 and described monkeys then
living on Mt. Misery (SMITH 1745:35, quoted in Sade and Hildrech 1965:68-69). However, Smith's description of the natural
history of Nevis does not mention monkeys on his home island,
and the earliest reference that I have found to monkeys on
Nevis is in COLERIDGE (1825:198).

In summary, then, all sources except Schomburgk's history
of Barbados and the 1642 deed to land at Apes Hill indicate
that feral populations of monkeys were established on St. Kitts
and Barbados between 1650 and 1670, and on Nevis between 1745
and 1825. No 17th or 18th century document that I have seen

contains a species name or an unambiguous description of the
species to which any of these animals belonged.

When did the migration end?
      Although the Barbadian bounty laws suggest that the impor-
tation of monkeys might have been prohibited after the animals
became pests, I have found no evidence to that effect.   The
sources that I examined include Rawlin (1699), Hall (1764), and
the 44-volume Calendar of State Papers, Colonial Series, America and the West Indies 1574 -1738.   All of these sources agree
with regard to the dates and contents of the Barbadian bounty
laws, but none of them refers to Barbadian import restrictions
on monkeys, and the Calendar of State Papers says nothing about
either bounty laws or import restrictions on monkeys in St.
Kitts or Nevis.   Furthermore, 17th to 19th century journals
kept by visitors and residents of Barbados, St. Kitts and Nevis
contain frequent comments on the Barbadian bounty laws, but no
comments at all on Kittitian bounty laws, or on laws against
importing monkeys to any of the islands (see Appendix, passim).
I am convinced that import restrictions did not exist as late
as 1738 on St. Kitts and Nevis and as late as 1762 on Barbados,
and that bounty laws did not exist on St. Kitts and Nevis as
late at 1738.
      The absence of import restrictions, in conjunction with
the importance of monkeys as pets and trade goods, suggest that
the monkey populations on Barbados, St. Kitts and Nevis could
have continued to receive frequent infusions of new animals for
more that two centuries after the commonly accepted termination
date of 1700.

When Did Green Monkeys Reach Barbados?

Schomburgk's Argument
      Although it is clear that feral monkeys were pests in Bar-
bados in the 1670's, their species is uncertain.   In his History of Barbados, SCHOMBURGK (1848:682-683) not only stated
that monkeys were present in Barbados when the first settlers
arrived, but also he stated that, from the outer appearance of
a living specimen that he had examined (he did not procure a
dead specimen), he considered the Barbadian monkeys of the
1840's to be "Cepus capucinus, Geoff., the Sai or Weeper, or a
very closely allied species (of New World monkey)" that was
nearly extinct in Barbados when he visited the island. This is
the earliest publication that I have found that contains a species name for Barbadian or Kittitian monkeys.
      ALLEN (1911:252-253), in a monograph on the mammals of the
West Indies, identified the Barbadian monkeys as Cercopithecus
sabaeus, and dismissed Schomburgk's report as follows:

"Schomburgk ... states that this monkey (viz.
Allen's C. sabaeus) was then nearly extinct, but
his belief may have been due to his misapprehension
that it was a native species of Cebus."

At best, Allen's statement is a complete non sequitur; at
worst, it suggests that Allen simply dismissed evidence that
was incompatible with his preconceived notions. A. H. Clark,
Allen's informant who had visited Barbados in 1903, identified
the animals as cercopithecines, so it followed that Schomburgk's 1848 statement that they were cebus was wrong.

As a result of Allen's dismissal of Schomburgk's work,
Schomburgk's statement that the Barbadian monkeys were cebus
has been virtually expurgated from the literature. Sade and
Hildrech (1965), Poirier (1972), McGuire (1974), and Ervin and
McGuire (1974) cited Allen's work, and therefore must have
known about Schomburgk's as well, but none of them even mentioned Schomburgk's assertion that the Barbadian monkey of the
1840's was Cebus capucinus or a closely related New World species. Perhaps as a result of the currently strong belief that
C. a. sabaeus is and always has been the only Barbadian
monkey, Poirier (1972:20-21) incorrectly stated that green monkeys were reported in Barbados as early as 1682, even though
the wording of the bounty laws from 1680 to 1714 provides no
evidence concerning the behavior or species of the Barbadian
monkeys.

If Schomburgk were just another of the many wayfarers who
visited Barbados and published journals of their travels, it
might be reasonable to dismiss his identification of the Barbadian monkeys as an inconsequential error. But Schomburgk was
not that kind of person. He was a skilled historian, a systematic biologist who made important contributions to the taxonomy
of South American flora, and a long-time resident of South
America and the West Indies. His History of Barbados

"... is a classic in West Indian history. It is a
scholarly study ... based largely on research done
in the British Museum Library and that of the
Library Society of Barbados. Much use has been
made of official publications, local newspapers,
and information gained by personal experience. The
work is an especially useful guide to, and reference for, the island's pre-1834 history" (Ragatz
1932:183, quoted with approval in Handler 1971:96).

If Schomburgk needed assistance in identifying the monkeys
that he saw, he could have used Linnaeus (1766), Buffon (1812,
1835) or Jardine (1833), to mention only a few relevant sources
that undoubtedly were available to him, certainly in London and
probably in Barbados. Jardine's color illustration of a green
monkey, and four of Buffon's black-and-white illustrations of
green monkeys and capucins are reproduced here in Figure 2.

FIGURE 2.   19th century monkey illustrations   and   original
captions.
    2.1.   C. sabaeus (green monkey);   Jardine 1833, Plate XIII.
    2.2.   Callitrix (green monkey);   Buffon 1812, Plate 394.
    2.3.   Sai or Weeper (cebus);   Buffon 1812, Plate 407.
    2.4.   Brown Capucin (cebus);   Buffon 1812, Plate 403.
    2.5.   Sai (cebus);   Buffon 1812, Plate 408.

It may be significant that Figure 2.3, which appears as Plate
407 in Buffon (1812), originally bore the caption "Sai or
Weeper", just as in Schomburgk's report of the Barbadian monkeys. It seems extremely unlikely to me that a person with
Schomburgk's background, skills and resources would have made
the kind of error that Allen attributed to him.

I have found no direct evidence concerning the identity of
Barbadian monkeys prior to Schombergk's report. The earliest
report that unambiguously identifies Barbadian monkeys as C. a.
sabaeus is in ELLIOT (1905:536-537), and the source of that report is A. H. Clark of Boston, the same person whom Allen
(1911) cited as his source. POCOCK (1907:726-727) reported
that C. sabaeus was abundant in Barbados, but cited no source
at all. The earliest unambiguous identification of the Kittitian monkeys is in SCLATER (1866:79-80) who does not mention
Barbados. It is followed by Elliot's (1905) and Allen's (1911)
reports that pertain to both islands. Brief descriptions of
Barbadian monkeys that appear in HUGHES (1750:66) and CHESTER
(1869:39) are insufficient to discriminate between green and
cebus monkeys, and Barbadian bounty laws describe neither the
appearance nor the behavior of the monkeys.

The circumstantial evidence offered by the 1642 deed to
land at Apes Hill is important here. If the hill derived its
name from a feral population of monkeys that lived there, the
monkeys may have been introduced by Amerindians during their
2000 years of intermittent occupation of Barbados prior to the
beginning of British colonization. In this case, it is virtually certain that the monkeys originated in the New World. Alter-
natively, the monkeys may have been introduced by the British
and become established in the interior of the island by the
mid-1630's. In this case, it is at least as likely that the
monkeys originated in the New World as in the Old World since
massive importation of slaves from Africa did not begin in
earnest until the establishment of the sugar plantation economy
in the West Indies in the 1640's.

It is by no means unreasonable to think that a cebus population could have become established in Barbados. Capucins are
easily tamed, make excellent pets, and were favorite organ
grinder monkeys (Hall and Kelson 1959:223). Various species of
cebus are found from Panama to Brazil, including parts of
Guyana and the island of Trinidad (Moynihan 1976:100-109).
Since the trade connection from Guyana and Trinidad, via Barbados, to England was one of the most important in the British
Caribbean, New World monkeys had many opportunities to reach
Barbados in large numbers, and the monkey that the Barbadian
sent to his friend in England in 1671 (PRO-CSPCS 1671) may simply be an example of this migration. Furthermore, cebus monkeys, like green monkeys, are notorious for raiding field crops
and plantations (Moynihan 1976:106, 109); hence the bounty
laws and early reports of monkeys in Barbados are no more compatible with the presence of green monkeys than with that of
cebus monkeys.

Based on recent fieldwork in Barbados, Horrocks (1984:32-33) reports that older Barbadians, especially in the Parish of St. John, still distinguish between the "common brown monkey" (C. a. sabaeus) and another kind of monkey that is "black and white", and they acknowledge that the latter is extremely rare. She suggests that the rare black-and-white monkeys, which she has not seen, could be C. a. tantalus, and that those monkeys could explain Schombergk's possible error. She argues that tantalus looks more like cebus than sabaeus does, since both cebus and tantalus have relatively shorter muzzles, darker pelages, and more conspicuous white browbands than does sabaeus. On the other hand, she notes that because of the differences just mentioned, cebus monkeys are more conspicuous that green monkeys. If cebus were present when Schomburgk wrote, monkey hunting could have resulted in their selective elimination without having had the same impact on the less conspicuous green monkeys.

Linguistic Clues

An intriguing suggestion concerning the identity of Barbadian monkeys prior to Schomburgk's book appears in a footnote to a poem entitled Barbadoes (CHAPMAN 1833:6, 88). Chapman uses the term "Jacko" to refer to Barbadian monkeys as COLERIDGE (1825:198) uses "Jacco" to refer to Kittitian monkeys. If the definition of "jacko" or "jacco" could be established, it might help to solve the problem posed by Schomburgk's report.

Although the Oxford English Dictionary (1971) has no entry for "jacko" or "jacco", it offers several hints concerning possible origins and meanings of the term. The first appears in the O.E.D.'s entry for "jackanapes", a term that first appeared in 1449 as "an opprobrious nickname of ... the Duke of Suffolk whose badge was a clog and chain, such as was attached to a tame ape." Shortly after his death in 1450, the Duke was "referred to as an ape, and entitled Jack Napes, this being inferrentially already a quasi-proper name for a tame ape." From 1526 onwards, "jackanapes" was used as a common noun meaning ape or monkey. Perhaps "jacko" is nothing but a short form of "jackanapes", but this explanation may be too simple.

Another possible derivation is suggested by the O.E.D.'s entry for "jocko". This term had its origin in 1625 in a work by Andrew Battell, in which engeco was given as the native name of the chimpanzee in Gabon, West Africa. Huxley (1898:19-20) reports that Buffon had access to a French translation of Battell's work, and in an attempt to systematize what was then known about gorillas and chimpanzees, published a chapter in 1776 entitled "Les Orangs-outangs ou le Pongo et le Jocko". Huxley argues that engeco thus became "jocko" and was spread worldwide in that form because of the popularity of Buffon's work. Eventually it came to be used as a familiar name of any monkey or ape; hence, Chapman's "jacko" may be a variant form of Buffon's "jocko". A third derivation that may or may not entail either or both of the first two is suggested by the

O.E.D.'s entries under "macaco", "macacus" and "macaque", all
of which have been applied to primates, each with a narrow and
a broad definition.   Their narrow definitions are mutually
exclusive, but their broad ones overlap so completely that in
the 18th and 19th centuries they were used interchangably as
another familiar name for monkeys.   For example, Ober (1880)
reported that residents of Grenada commonly used "macaque" to
refer to monkeys, but noted that it was a familiar term in the
French patois, not a species name in the restricted sense.
Under "macaco", the O.E.D. includes the following quotation:

>1874 Slang Dict., Murkarker, a monkey, vulgar Cock-
>ney pronunciation of Macauco.   Jacko Macauco ...
>was the name of a famous fighting monkey, who used
>nearly fifty years ago to display his prowess at
>the Westminster Pit.

"Jacko Macauco" may be an example of rhyming slang that originated in the Cockney underworld early in the 19th century
(Fowler 1965:525).   Perhaps the rhyme resulted from combining
"jocko" (from Battell through Buffon) with "macauco", but a
less pedantic hypothesis is that "jacko" (perhaps from "jackanapes") was attached to "macauco" because of the obvious rhyme,
and "jacko" became widely known because of its catchy attach-
ment to "macauco". But at least one more possibility remains.
     The fourth hypothesis rests on the great European popular-
ity of green monkeys from St. Jago.   In appearance and sound,
"Jago" and "jacko" - as well as "jocko" - closely resemble each
other.   Hence "jacko" may be an Anglicized form of "Jago". It
is then possible that "jackanapes", at least in its fully developed form in the 17th and 18th centuries, was a portmanteau
word encompassing "jocko" from Buffon, "jackanapes" from the
Duke of Suffolk, and "Jago apes" from the Cape Verde Islands.
The "jackanapes" entry in the O.E.D. contains two quotations
that may tentatively support this hypothesis:

>1698 Fryer, Acc. E. India & P. 7. Some brought
>Jackanaps's, such green ones are are commonly seen
>in England to be sold.

>1688 R. Holme, Armoury VII, 70/2. The Jack-an-
>Apes on Horseback, or the fantastic Cowslip, hath
>the flower all green and jagged.

That the first quotation refers to a kind of monkey and
the second to a kind of flower may be less important than the
fact that both are described as green, the color of the St.
Jago monkey.   If "jacko" is even in part an Anglicization of
"Jago", it may have had both the general meaning of "monkey"
and the specific meaning of "green monkey" in 19th century
reports from the West Indies. In this regard, it would be useful to know whether Jacko Macauco, the famous fighting monkey,

was green.
     If it could be established that Chapman used "jacko" in
the narrow sense of "green monkey" rather than as a familiar or
poetic term for all monkeys, we would have independent evidence
of an error of omission or commission by Schombergk. Such
evidence remains elusive, and I am unable to solve the problem.

Summary
     If Schomburgk correctly identified the Barbadian monkeys
as cebus in 1848, it is likely that the bounty laws pertained
to cebus monkeys that were established on the island when the
first British settlers arrived or that became established on
the island shortly thereafter, and that probably became extinct
sometime between 1848 and 1903. If that happened, it follows
that green monkeys coexisted with cebus monkeys in Barbados
prior to 1848 and nobody ever commented on the presence of two
monkey species on the island, or that green monkeys were not
introduced until the last half of the 19th century. Until this
problem is solved, it is unsafe to assume that feral populations of African monkeys inhabited Barbados prior to 1848.
     I have found no information that specifically challenges
the view that the monkeys that Labat saw in St. Kitts in 1700
were cercopithecines.

Alternative Interpretations

     If the monkeys of Barbados, St. Kitts and Nevis consti-
tuted a single population, answers to questions about migra-
tions to any one island should apply directly to the others.
An assumption to that effect is implicit in the recent literature on the West Indian cercopithecines, but there are several
alternative interpretations that are more reasonable.
     Even if C. aethiops were the only species of monkey that
had ever lived in a feral state on Barbados, St. Kitts and Nevis, and even if those populations became established in the
17th century, the following are among the more obvious alternative historical scenarios.
     a) The French introduced the monkeys to St. Kitts from
Senegal or Gambia, and Kittitian monkeys were introduced subsequently to Nevis, Sint Eustatius, and Barbados.
     b) The French introduced the monkeys to St. Kitts from
Senegal or Gambia, and Kittitian monkeys were introduced subsequently to Nevis and Sint Eustatius. The British introduced
the monkeys to Barbados from the Gold Coast.
     c) The French introduced the monkeys to St. Kitts from
Senegal or Gambia, and the British introduced them to Barbados
from the Cape Verde Islands where they had immigrated from the
adjacent coast of Africa in the 15th or 16th century. The monkeys on Nevis probably came from St. Kitts, but could have come
from Barbados. Those on Sint Eustatius almost certainly migrated from St. Kitts in the 20th century.

d) Both Labat and Schomburgk were wrong.   The British introduced the monkeys to Barbados, St. Kitts and Nevis from the Gold Coast.   There could have been simultaneous or independent introductions to all three islands, or sequential introductions in which the monkeys used the islands as stepping stones.

These four relatively simple possibilities suggest that the West Indian islands may contain monkey populations whose histories are strikingly different from the generally accepted scheme.   If the founders of one of the West Indian populations came from the Cape Verdes, where they could have been isolated from African populations for two centuries prior to their migration to the Caribbean, then their current period of isolation from Africa may be 500 years rather than 300. If the Kittitian population originated in Senegal and the Barbadian on the Gold Coast, those two populations may have been separated from Africa for 300 years but they also originated over a thousand miles apart at opposite ends of the region inhabited by C. a. sabaeus.   And their genetic separation could be even greater if one group migrated from Senegal via the Cape Verdes and the other from the Gold Coast.   If one of these suggestions is correct, then populations of green monkeys in the West Indies may represent a much more fascinating and complex natural experiment in evolutionary differentiation than has been suggested by earlier accounts of the animals' histories.

On the other hand, if most or all of the island populations received frequent and major infusions of sabaeus from Senegal, the Cape Verdes and the Gold Coast, and perhaps tantalus and other subspecies from other parts of Africa, and if those infusions continued from the mid-17th to the mid-19th century, then the West Indian populations may represent an equally fascinating but very different natural experiment that is based as much on interbreeding as on genetic isolation.

Finally, if Schomburgk was right and cebus was the only species of monkeys in Barbados until sometime between 1848 and 1903, the green monkeys now living in Barbados separated from their parent stock, in either Africa or St. Kitts, only a century ago.

Although the generally accepted account of cercopithecine migration to the West Indies is simple and appealing, I think it is one of the more improbable interpretations that can be offered concerning the origins of the West Indian populations. If further work is to be undertaken with these animals, and if their potential value as subjects of natural experiments is to be realized, the more complex but realistic scenarios must be examined in detail.   The ones that I have suggested are not necessarily the best, and better ones may emerge as more historical data becomes available.   But it is unlikely that historical data alone will ever yield a clear solution to the problems raised here.

The historical sequences that I have called "alternative scenarios" can be rephrased as competing hypotheses that can be tested by examining the genetic characteristics of the animals

of all of the relevant islands in the West Indies, the Cape
Verde Islands, and various locations along the west coast of
Africa from Senegal to Angola. Clearly the task will be a large
one, but if further work with the West Indian animals is to
take their origins and history into consideration, the problems
raised here must be solved.

III. POPULATION CHANGES AMONG BARBADIAN MONKEYS

Introduction

The kinds of problems that beset attempts to reconstruct the migration history of West Indian green monkeys wherever they may be found are not reduced by focusing on a problem that is tightly circumscribed geographically. This chapter explores in depth the population history of Barbadian monkeys, and once again questions are far easier to ask than to answer.

The few remarks about Barbadian monkeys that have appeared in the primate literature have created the impression that C. aethiops arrived in Barbados about 1650, multiplied rapidly, became an important agricultural pest by 1682, then virtually disappeared from the island shortly thereafter as a result of bounty hunting. Accordingly, the only monkeys that are there now are supposed to be descendants of the early population who managed to survive the "great extirpation".

The long, narrow population bottleneck that is implied by the commonly accepted account could have dramatically accelerated rates of evolutionary change by selecting strongly for exceptional phenotypes. On the other hand, since green monkeys and their congeners are known to be diverse and adaptable, a population bottleneck might be of minor importance even if it occurred. In either event, we need to know whether a bottleneck actually occurred, and if so, what its nature and dimensions were.

I begin the chapter with overviews of the physical environment and social history of Barbados, then review historical and recent data concerning numbers and distribution of Barbadian monkeys. Next I attempt to interpret the population data in the context of the island's natural and social history. I conclude the chapter by presenting three competing hypotheses concerning the population history of Barbadian monkeys. Some material that appears in this chapter was discussed from different perspectives in Chapter II, but most of it makes its first appearance here.

The Setting

Geology.

Barbados is located in the North Atlantic Ocean on the eastern perimeter of the Caribbean Sea at 13deg10min N. Lat. and 59deg35min W. Long. It is approximately 34 km. north to south and 23 km. east to west, and has an area of 430 sq. km. Its highest elevation is 340 meters.

Unlike St. Kitts, Nevis and its near neighbors in the Antilles, Barbados is not of volcanic origin. Eighty-seven percent (373 sq. km.) of its surface is covered with a thin cap of
Pleistocene limestone, while the remaining 13% (57 sq. km.) is
covered by tertiary sediments that also underlie the limestone.

The island is located over 150 km. east of the Antillean
Volcanic Arc. The westward spreading Atlantic sea floor, which
is descending under the Caribbean tectonic plate, has slowly
forced the island upward as the topmost peak of the Barbados
Ridge, an accretionary wedge similar to those that have produced islands such as Aruba, Bonaire and Curacau in the Nether-
lands Antilles and Timor in Indonesia.

The upper surface of Barbados reached sea level about
700,000 years ago. Relatively slow uplift continued throughout
the Pleistocene, and fringing coral reefs developed more or
less continuously throughout that period. The reefs are preserved now as limestone terraces, up to eleven of which are
discernable at some places on the west side of the island. The
events that produced the alternating terrace-and-cliff structure remain obscure.

The focus of uplift is a point off the northeast coast of
the island, and that also is the direction from which tradewinds hit the island. As a result of uplift fracturing of the
limestone combined with severe weathering, the limestone cap is
missing from the Scotland District. There, highly eroded sedi-
mentary formations appear as steep, rugged hills that reach
from the crest of the island to the northeast coastline.

Figure 3 shows the location of the limestone cap and the
Scotland District, and a section through Mt. Hillaby, the
highest point on the island. In comparison with St. Kitts and
Nevis, Barbados is relatively low and flat.

Geological features that provide most of the habitats for
Barbadian monkeys are fracture and erosional gullies that occur
on the highest parts of the limestone cap as well as at inter-
vals of several hundred yards along the face of each terrace
and on the steep hillsides in the Scotland District. Some gullies are short and shallow, being no more than minor clefts in
rims of terraces. Others extend for 10 or 11 kilometers, beginning near the crest of the island and winding through limestone
cracks and erosional channels all the way to the coast. As
longer gullies cross limestone terraces, they pick up tributar-
ies and become shallow and relatively broad. At the edges of
terraces they descend rapidly through narrow channels with
steep walls up to 50 meters high in some places. Three of the
gullies contain permanently flowing but tiny streams.

In addition to gullies, the soluble limestone cap contains
many sinkholes and caves, including Harrison's Cave which has
been illuminated internally and opened for guided tours. Also,
when the island was younger, coastal wave action seems to have
eroded caves in cliff faces and gully walls. Residents of the
island assert that monkeys use shallow caves in the vertical
walls of gullies and cliffs as shelters and sleeping sites, and

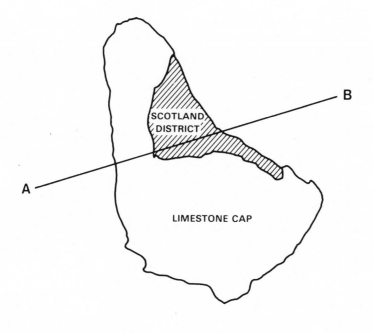

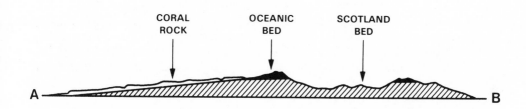

FIGURE 3. Geology of Barbados

I have photographed the openings of shallow caves where monkeys have carried sugarcane from fields to eat in these safer locations.

Geological materials upon which these comments are based include Jukes-Brown and Harrison (1891), Trechman (1925, 1937), Senn (1940), Daviess (1971), James, Mountjoy and Omura (1971), Poole and Barker (1983), and R. Speed (pc).

Climate.

Barbados lies within the belt of the easterly tradewinds that traverse the southern part of the North Atlantic Ocean. In general, the climate displays only slight seasonality in rainfall and temperature.

In lowland coastal districts, the dry season is from mid-December to mid-May, and the wet season from mid-May to mid-December. The dry north and south coasts receive about 100 cm. of rain per year, most of which falls during the wet season. Rainfall seasonality virtually disappears with increasing elevation, and the central highlands receive an annual total of about 200 cm. distributed evenly throughout the year. Spatial and seasonal distributions of rainfall are depicted in Figure 4, and mid-elevation rainfall averages by month from 1847 through 1960 appear in Table I.

MONTH	MEAN MONTHLY RAINFALL (IN CM.)	MONTH	MEAN MONTHLY RAINFALL (IN CM.)
January	8.56	July	15.90
February	5.59	August	18.57
March	4.90	September	19.43
April	5.94	October	20.04
May	8.15	November	19.66
June	13.74	December	12.16

MEAN ANNUAL RAINFALL (IN CM.)   152.64

TABLE I. Mean monthly rainfall at Codrington Agricultural Station, St. Michael, Barbados, 1847-1960 (adapted from Skeete 1961).

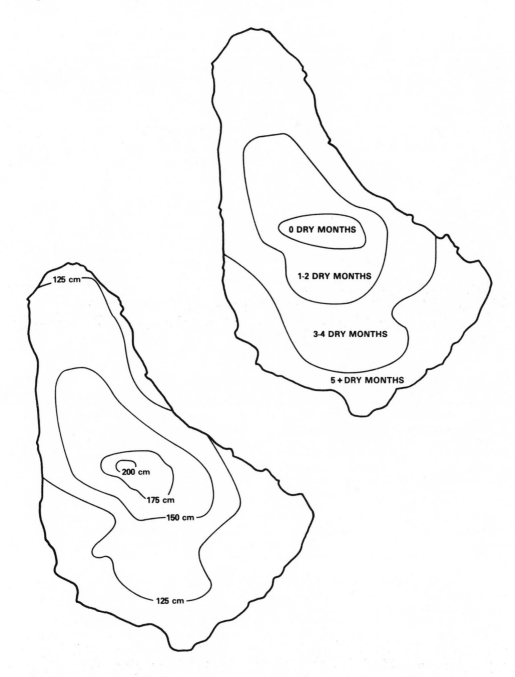

FIGURE 4.   Rainfall distribution in Barbados.

The warm season is May through September and the cool season is October through April. Temperature rarely exceeds 33deg C. or falls below 16deg C. in any part of the island, and is usually 2-3deg C. lower in the central highlands than in coastal districts.   Monthly averages from 1903 through 1960 appear in Table II.

MONTH	MEAN TEMP. (C.) HIGH	LOW	MONTH	MEAN TEMP. (C.) HIGH	LOW
January	28.3	21.1	July	30.0	23.3
February	28.5	20.9	August	30.2	23.2
March	29.2	21.3	September	30.2	23.0
April	29.8	22.1	October	29.8	22.8
May	30.4	23.0	November	29.3	22.6
June	30.2	23.4	December	28.6	21.8

TABLE II.   Mean maximum and minimum temperatures, Codrington Agricultural Station, St. Michael, Barbados, 1903-1960 (adapted from Skeete 1961).

Tradewinds blow at an average speed of 10-20 km./hr. year-round.   Calm days, and days with average wind speeds greater than 30 km./hr., are rare and isolated.   Wind noise is ever-present.

Barbados lies to the south of the usual Caribbean tropical storm track, and its weather is affected only occasionally and moderately by hurricanes.   Since the British settled there in 1627, the island has been hit by them only five times (1675, 1780, 1831, 1898, and 1955), an average of once every seventy years.

Meteorological records for Barbados are exceptionally good in both quantity and quality.   Information summarized above comes from Skeete (1961) and Gooding (1974), and detailed records and other useful documents are available on the island.

Indigenous Flora and Fauna.
Because of its location far to the east of the Antillian Island Arc, its low maximum elevation and its peculiar geology, Barbados had a climax vegetation that was distinctly different from that of most islands of the Lesser Antilles when the British settled there in 1627.   Northern and southeastern coastal districts of porous limestone, receiving only 100 cm. of rainfall annually and exposed to nearly continuous tradewinds, were covered with xerophytic woodlands and dense scrub.   At higher elevations, conditions within gullies - especially persistent moisture from rainfall accumulation and seepage from the lime-

stone, as well as shelter from strong winds - permitted the
gullies to develop lush mesophytic woodlands and rain forests
with dense forest canopies, shrubs of intermediate height,
ground stories of herbs and ferns, and an abundance of climb-
ers. The remainder of the island, receiving moderate to heavy
rainfall and at least partially protected from tradewinds and
saltwater spray, was covered with dense mesophytic or xerophytic woodlands (Watts 1966:34-37). Unlike St. Kitts and
Nevis, which in part occupy "mountain" and "high mountain"
zones in Beard's (1949) classification of Lesser Antillean cli-
mates and vegetation, all of Barbados lies in Beard's "dry" and
"middle" zones (Watts 1966:17-18).
    The indigenous land fauna of Barbados was severely
impoverished. According to Ligon (1657), Hughes (1750) and
other early observers, it included bats, a few species of
birds, a nonpoisonous snake, lizards and insects. Schomburgk's
(1848) assertion that monkeys were present in 1627 in plausible
but unsubstantiated.
    This brief overview of indigenous flora and fauna comes
from the sources cited above, as well as Gooding, Loveless and
Proctor (1965), Gooding (1974), and Vernon and Carroll (1965).

A Monkey's Eye View of Barbadian History

    During the first two decades after settlers reached Barba-
dos, clearing the woods for subsistence farming proceeded slowly. But when sugarcane was introduced and immediately became a
successful cash crop late in the 1640's, sugar and slaves became the twin bases of the economy, and recession of the forest
edges accelerated rapidly. Watts (1966:45), in a monograph devoted specifically to the history of Barbadian flora, argues
that virtually all of the island's arable land had been cleared
by 1665. The mesophytic woodland was replaced by sugarcane,
ground crops for human consumption, corn for consumption by
humans and livestock, and pastureland. The xerophytic scrub was
replaced by arid pastureland interspersed with cacti. Turner's
Hall Wood, a permanently protected fifty-acre remnant of climax
forest located on a steep hillside at an elevation of 180 - 240
meters in the Scotland District, is the only woodland in Barbados that has retained any of its 17th century floral diversity.
    The clearing of agricultural lands, the establishment of a
large slave population, and the production of sugar had many
implications for feral monkeys, the most obvious of which was
the elimination of arboreal habitats on arable lands. That
problem was intensified greatly by the denudation of gullies
and the bases of cliffs. In addition to turning the larger
trees into lumber, the households and sugar factories rapidly
consumed other wood as fuel. "Plantation woodlots were cut
down both for fuel and to increase the acreage of cane" (Watts
1966:45), and maps and descriptions of land usage on 18th and
19th century Barbadian plantations do not mention woodlots

(Handler and Lange 1978:43-67;   Starkey 1939:111).   As early as
1667, an observer reported:

> "At the Barbadoes all the trees are destroyed, so
> that wanting wood to boyle their sugar, they are
> forced to send for coales from England" (PRO 1667,
> quoted in Watts 1966:45).

When that strategy proved uneconomical, wood was imported from
New England, the Maritime Provinces of Canada, Guyana and other
West Indian islands (Schomburgk 1848:156-159).   According to
Edwards (1807:146):

> "Barbadoes ... is dotted with houses as thick as on
> the declivities in the neighborhood of London or
> Bristol, but with no woods, and with very few
> trees, even on the summits of the hills;   two or
> three struggling cocoas near each dwelling-house
> were all the trees to be seen."

    Turner's Hall Wood was guarded by rangers who permitted
underbrush to be cleared but protected larger trees.   Theft of
firewood was punishable by imprisonment until the 20th century
(Anon. 1897).
    In contrast to the rather grim picture that Watts pre-
sents, Handler (pc) and Campbell (pc) suggest that the deforestation of Barbados was extreme only in comparison with either
its pre-1665 condition or the densely forested volcanic islands
elsewhere in the West Indies, and that many plantations retained wooded areas after 1665.   If Handler and Campbell are
correct in arguing that Watts exaggerated the extent to which
the woods were flattened, it may be more accurate to say that
Barbadian woods were rare rather than absent after 1665.
    Except for orchards that settlers planted, the relatively
or absolutely treeless condition of Barbados remained unchanged
until the 1950's when bottled gas, kerosene, heating oil and
electricity began to replace firewood and charcoal as fuel.
During the last quarter-century, large stands of trees have returned to agriculturally useless lands and many gullies now are
filled with dense woods.   The woods that appear at the bases of
inland cliffs are, in many cases, extensions of the gully woodlands, but they tend to be considerably drier, their trees are
lower, and their shrubs often form impenetrable thickets.   At
higher elevations the woods appear as more mature second-growth
forests.   The lower storey has returned to Turner's Hall Wood,
and except for the presence of several introduced species, it
probably looks much the same now as it did in 1627.   Welchman
Hall Gully and Farley Hill, two small national parks, are heavily wooded now, as is much of the upper rim of the Scotland
District.

Another major consequence of the establishment of sugar plantations was the regular production of introduced domesticated foods that were palatable to both humans and monkeys. By 1676, orchards, fields and gardens all over the island yielded sugarcane, cassava, yams, corn, plantains, bananas, pineapples, melons, potatoes, turnips, figs, dates, grains, legumes, and herbs. By 1700, the list included mangoes, tomatoes, okra, eddoes, sweet potatoes, groundnuts, avocadoes and pumpkins (Watts 1966:88-92).

Because of the heavy emphasis on sugar production, the island has never yielded enough provisions to support all of its people (Schomburgk 1848:156-169; Starkey 1939 passim; Handler and Lange 1978:86-90), but its dependence on imported foods has varied from time to time. Both Handler and Lange (1978:89) and Starkey (1939:74) assert that the island was almost totally dependent upon imported foods between 1680 and 1700. That degree of dependence became economically intolerable early in the 18th century and did not recur until the "sugar boom" during the First World War (1913-18). Currently there are market gardens and small private gardens throughout the island, and it is a legal requirement that 12.5% of each plantation be used to raise provisions.

In addition to cultigens and wild plants, settlers introduced domesticated animals including dogs, cats, chickens, African black-bellied sheep, camels (now extinct on the island), asses, horses, and cows, the latter being valued equally for their dairy products and their manure for fertilizing cane fields. The earliest domesticated animals on the island probably were hogs, whose introduction is traditionally attributed to Pedro a Campos, the Portuguese explorer who discovered Barbados in 1536 and left the animals there to breed as food for future sailors (John Scott, n.d., cited in Harlow 1961:1). Settlers also introduced two species of rats (R. rattus and R. norvegicus), the house mouse (M. musculus), the mongoose (Herpestes javanicus) which was imported in 1878 to control rats and rapidly became a pest in its own right, a raccoon (Procyon gloveralleni), and monkeys. The only predators against Barbadian monkeys are humans and dogs (see Schomburgk 1848:683 concerning wild dogs that preyed on Barbadian sheep).

By 1680 the human population of the island reached 70,000. Despite massive importation of slaves, its size remained constant at approximately 75,000 until the slave trade ended in 1807. As slave owners began to provide better care for their slaves, the population grew, and at emancipation in 1834 it stood at 104,000. By 1900 it had grown to 200,000. Intensive immigration and a series of epidemics reduced it to 157,000 by 1921, but by 1970 it had soared to 243,000 where it seems to have stabilized again (Table III). With more than 570 people per square kilometer, Barbados now is one of the most densely populated countries in the world (Lowenthal 1957, Rickards 1977; Handler and Lange 1978).

Prior to 1834 the vast majority of Barbadians lived in approximately 400 slave villages located near managers' and overseers' houses on plantations all over the island (Handler and Lange 1978:37). At emancipation, many of the slave villages were eliminated and tenantries - functionally and structurally equivalent to slave villages, but occupied by freedmen - were established on the margins of plantations. The island now has four major towns (Bridgetown, Holetown, Speightstown and Oistins), several villages, and more than 600 hamlets most of which are or were tenantries, and about 1200 km. of paved roads.

YEAR	POPULATION	SOURCE
1627	0	-
1680	70,000	Lowenthal (1957)
1715	69,500	Handler and Lange (1978)
1757	88,000	Handler and Lange (1978)
1800	77,000	Lowenthal (1957)
1802	82,700	Handler and Lange (1978)
1823	99,700	Handler and Lange (1978)
1834	104,300	Handler and Lange (1978)
1844	122,000	Lowenthal (1957)
1900	200,000	Lowenthal (1957)
1921	157,000	Lowenthal (1957)
1951	210,000	Lowenthal (1957)
1970	243,600	Rickards (1977)
1980	246,082	Barbados Stat. Service (pc)

TABLE III. Estimated human population of Barbados, 1627-1980.

LAND USE	AREA	
	SQ. KM.	PCT. OF TOTAL
Sugarcane	197	46
Food crops	80	19
Permanent pasturage	49	11
Other	104	24
TOTAL	430	100

TABLE IV. Land use in Barbados in 1977 (adapted from Rickards 1977).

During the last 300 years, land use patterns have not changed markedly from those that were established by 1665 (see Starkey 1939:61,74,111,119 and Handler and Lange 1978:65 for quantitative data concerning land use on single plantations in the years 1650, 1685, 1796, 1811 and 1842). About 35-50% of the island's 430 sq.km. has been planted in sugarcane each year, 10-15% has been used as permanent pasturage, and 20-40% has been used to raise provisions. The remainder - about 20% - has been occupied by towns, villages, hamlets, roads, and agriculturally useless lands such as gullies, bases of cliffs, and eroded hillsides in the Scotland District. Land use in 1977 is summarized in Table IV.

Summary of Setting and History

Presumably most or all of the factors introduced above have affected the lives of the Barbadian monkeys. The island's peculiar geology provides habitats in gullies throughout the island. The higher and more evenly distributed rainfall in the highlands makes those areas more hospitable than the arid north and south coastal regions. Although major hurricanes have hit the island five times since 1627, their impact on primate populations is unknown. Except for humans, dogs are the monkeys' only predators.

Human impact on the Barbadian landscape has been profound. The island lost most of its trees by 1665; hence much of the natural habitat of arboreal or semi-arboreal Cercopithecus and Cebus species apparently was destroyed and did not regenerate until the 1950's. From 1665 onwards, however, sugarcane and provisions were planted all over the island, and Cercopithecus and Cebus species eat both. Although land use patterns have been relatively stable throughout the island since 1665, the human population has become extremely dense, especially along the west and southwest coasts. Together the environmental characteristics introduced here provide the context in which to examine primate populations in Barbados.

Barbadian Monkey Population Data

Early Reports.
The earliest direct evidence concerning primate populations in Barbados is provided by the bounty laws of 1680 - 1714. Although they say nothing about the appearance or behavior of the monkeys, the laws of 1680 and 1682 both state that the number of monkeys had increased in recent years. Furthermore, the mere fact that the laws were passed repeatedly suggests that monkeys were causing a great deal of trouble. Unfortunately, the nature of the trouble is unspecified.

The next reference that I have found to Barbadian monkeys appears in HUGHES (1750:66). In this major work on the natural history of Barbados, monkeys appear in four sentences. Hughes said that they were not very numerous; lived in inaccessible gullies; raided fruit trees, yam and potato gardens, and canefields; and because of the bounty, "they yearly rather decrease than multiply".

In a didactic poem characterized by artistic license and hyperbole, and imitative of GRAINGER's (1764) poem from St. Kitts, CHAPMAN (1833:5-6,88) devoted one verse and a footnote to Barbadian monkeys. He said the monkeys had been driven from their "ancestral wood" and "exterminated" by man in "the empire that was once (the monkey's) own". Furthermore, he noted that the slaves admired the sagacity of the monkeys, but that their admiration had not kept them from killing monkeys on whose heads there was a bounty.

CHESTER (1869:39,49) noted that Barbadian monkeys were confined to a few gullies and the undercliff of the Scotland District, and that "advancing cultivation and rewards for their extirpation" had rendered them extremely rare.

Shortly after the beginning of the 20th century, reports became more frequent. POCOCK (1907:727) cited no source when he said that C. aethiops was "still abundant at least in Barbados". ELLIOT (1905:536-537), citing A. H. Clark as his source, noted that C. aethiops was "common in certain districts of Barbados". ALLEN (1911:252-253), citing the same A. H. Clark, said that C. aethiops was found at only a few points in Barbados because of the almost complete deforestation of the island. He reported a group with half a dozen members at Foster Hall Estate, said they were common along the upper reaches of Joe's River, and commented on their destructiveness of fruits, vegetable gardens, and sweet potato patches. Finally, he ignored Hughes' and Schomburgk's references to bounty laws, but inferred from their other comments that monkeys had never been able to increase greatly in Barbados.

In a travelers' guide to the West Indies, OBER (1913:418-419) reported monkeys living in Turner's Hall Wood and warned hunters that they were difficult to obtain "for at one time in the past a bounty was placed upon their heads, and they came near being extirpated."

A team of nineteen zoologists spent several months in Barbados and Antigua in 1919, but according to NUTTING (1920:66) they paid practically no attention to land vertebrates in Barbados. Although they sighted monkeys "in a small patch of woodland in the northern part of the island", they did not secure specimens for identification.

ASPINALL (1927:23; 1931:104,106), a writer of guidebooks and popular histories of the West Indies, reported monkeys at Welchman Hall, in a mahogany grove at Porter's Estate near Holetown, and at Nicolas Abbey estate on the Northern Rim of the Scotland District. He recounted an anecdote about monkeys playing tennis on the court at Porter's Estate, but gave no useful information about them.

In 1978, my wife, Nancy Hubley, interviewed several Barba-

dians who migrated to Cape Breton Island, Nova Scotia, between
1910 and 1925. Although her own research focused on economic
history and trade connections between Barbados and Nova Scotia,
she inquired about people's encounters with monkeys before they
left Barbados. Some did not recall monkeys, but six stated
that when they were children or young adults, monkeys lived
near their homes at Speightstown, Hillaby, St. Simon, Black
Rock, Hall's Village, and Fairy Valley. The person from Fairy
Valley also remembered seeing monkeys at Lowther's Estate.

Chester, Allen, Ober, Nutting, Aspinall, and the Barbadian
immigrants to Nova Scotia state unequivocally that monkeys
lived at thirteen specific points and two general areas in Barbados during the period 1869 - 1931. The locations that they
mentioned are plotted on the map in Figure 5.

Taylor's Questionnaire, 1964.
In 1964, K. D. Taylor, an infestation control officer with
the Ministry of Agriculture in London, went to Barbados to assist in planning a program to eradicate rats and monkeys that
were destroying sugarcane. Rats were a longstanding problem,
but monkeys had only recently become a matter of serious con-
cern. Since Taylor's (1966) report, the only serious scientific publication that dealt with Barbadian monkeys before the
end of the 1970's, is not generally available, I paraphrase and
quote it extensively here.

In Barbados in 1964 there were about 300 plantations of
ten or more acres each, about 3,900 small holdings of one to
ten acres each, and about 23,700 small holdings of less than
one acre each (Handler, pc). As part of his project, Taylor
distributed questionnaires to 200 plantations but no small
holdings, and received replies from 64 of them. His report
does not identify plantations by name.

Fifty-four of Taylor's respondents reported that monkeys
were present occasionally or continuously on their plantations,
47 gave estimates of the number of monkeys seen regularly on
their plantations, 42 indicated that the numbers were increasing, and "several managers in the parishes of St. Phillip and
Christchurch reported monkeys where there had been none a few
years previously" (Taylor 1966:26). The report says nothing
about the number of monkey troops on each plantation, but it
notes that monkeys were generally seen in groups of two to five
individuals, and that there were a few reports of groups containing 70-100 animals. The 47 estimates of monkey numbers
that Taylor reported are plotted on the map in Figure 6.

Although 24 of his respondents indicated that crops had
been damaged by monkeys and nine estimated the damage to be
worth $1000 (Barbadian) or more per year, data from the island
as a whole indicated that monkeys were minor agricultural pests
in comparison with rats. Tacit support for that inference was
provided shortly thereafter by a publication of the Barbados
Ministry of Agriculture (Ingersent, et. al. 1969) which omitted
monkeys from a lengthy list of pests in vegetable gardens.

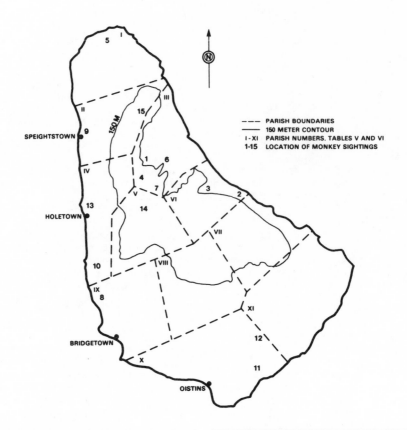

LOCATION	DATE OF SIGHTING	PLACE NAME	SOURCE
1	1869	Undercliff of Scotland District	CHESTER, 1869
2	1903	Foster Hall	ALLEN, 1911
3	1903	Joe's River	ALLEN, 1911
4	1913	Turner's Hall Wood	OPIF, 1913
5	1919	Northern Part of the Island	NUTTING, 1920
6	1917±7 yr	St. Simon	Hubley, 1978
7	1917±7 yr	Hillaby	Hubley, 1978
8	1917±7 yr	Black Rock	Hubley, 1978

LOCATION	DATE OF SIGHTING	PLACE NAME	SOURCE
9	1917±7 yr	Speightstown	Hubley, 1978
10	1917±7 yr	Hall's Village	Hubley, 1978
11	1917±7 yr	Fairey Valley	Hubley, 1978
12	1917±7 yr	Lowther's Estate	Hubley, 1978
13	1927	Porter's Hall	ASPINALL, 1927
14	1927	Welchman Hall	Aspinall, 1931
15	1927	Nicholas Abbey	Aspinall, 1931

FIGURE 5. Locations of monkeys in Barbados, 1869-1931.

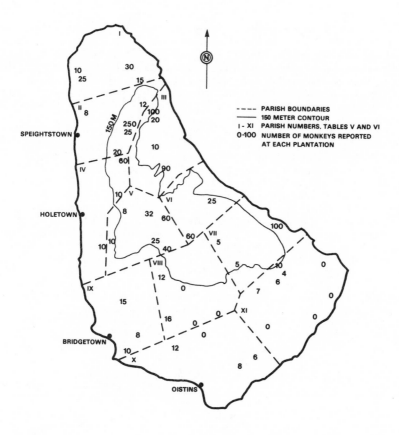

FIGURE 6.   Monkeys in Barbados in 1964.

Taylor reviewed current pest control techniques and noted that poison was ineffective because monkeys refuse to eat food that had been tampered with or left on the ground as bait, and because people in Barbados were ethically opposed to poisoning monkeys. Although he found weaker ethical objections to shooting monkeys, he noted that shooting, like poisoning, was an ineffective pest control technique. Most plantation managers did not take monkey shooting seriously and did not possess the best equipment for the job. Furthermore, "The effect of shooting is to deter rather than control since, on the whole, few monkeys are killed. Shots fired at a troupe seem generally to keep them away for a week or two but during this period they probably feed on neighboring plantations and the overall effect is therefore small" (Taylor 1966:26).

In an attempt to devise a better pest control technique, Taylor helped to organize and conduct an hilariously ineffective "monkey drive". The objective was to drive monkeys out of a gully and shoot them, primarily to demonstrate that the technique could be used in Barbados just as it had been in West Africa. The organizers chose Jack-in-the-Box Gully, one of the largest, most densely wooded and most precipitous gullies on the island, as the site for the experiment. Sixteen men with six shotguns, a portable public address system, three dogs, and hundreds of firecrackers and other noisemakers spent an entire afternoon trying to drive monkeys out of the gully. Monkeys were seen in the gully shortly before the drive began and again forty-five minutes after it ended, but none fled from the gully nor were shot in it during the drive. Apparently the monkeys simply hid in dense brush and in brush-concealed caves in the gully walls until the people and noises went away. As Taylor noted somewhat wryly, the full-scale test of the technique demonstrated its inapplicability to Barbadian conditions. Although it might work in small open gullies, it clearly did not work in one of the locations where a large monkey population was known to live.

Taylor concluded that monkeys were minor pests in Barbados in 1965 and that poisoning and shooting were ineffective techniques for controlling them. He predicted that the primate population would continue to grow as reforestation of gullies proceeded during the next two decades, and that a long term program of waste land utilization that would deprive the monkeys of woodland habitats in gullies would be the only effective way to solve the monkey problem.

Field Observations in 1977-78.

Taylor's prediction that the monkey problem would intensify was accurate. By 1975 the problem had become so serious that the 17th century bounty law was revived, responsibility for administering it was transferred to the Barbados Ministry of Agriculture, and the bounty was changed from five shillings to five dollars (P. Jeffers, Barbados Ministry of Agriculture, pc). Articles about monkeys began to appear in the popular press (e.g., Hutton 1977:48-49), and the monkeys' importance as agricultural pests became a common topic of conversation.

Two sets of population data were collected in 1977-78, the first by T. DuBois of Rutgers University and the second by Hubley and me. Both appear in Table V. DuBois inquired about or briefly observed a great many troops late in 1977 while searching for two good places to conduct a comparative study of mother-infant interactions among Barbadian monkeys. Her data for Hopefield Estate and Oughterson Gully are the result of repeated counts of one troop at each location, but the other observations and reports were not double-checked or cross-checked and must be interpreted as minimum sizes or order-of-magnitude estimates. DuBois made no attempt to survey the island systematically or completely, and she terminated her survey when she found two sites that met her research requirements. She lived on the southeast coast of the island, and her survey dealt mainly with the southern half of the island.

Hubley and I obtained our data through interviews and observations similar to those conducted by DuBois. In May and June 1978, we repeatedly observed the Witchwood Gully and Farley Hill troops and are confident that our reports on troop size and composition are correct for them. The other observational data are unchecked and represent minimum troop sizes. Since we lived on the northern section of the west coast, our survey dealt mainly with the northern half of the island.

Although the data that Taylor, DuBois, Hubley and I recorded do not represent a thorough survey of the Barbadian monkey population, they can be used cautiously to make some preliminary generalizations about population distribution and density, and about troop sizes and compositions. In Table V, site elevations appear with troop locations and sizes. In Table VI, the number of monkeys reported to Taylor from each parish appears with the number of monkeys reported in those parishes in the 1977-78 surveys. Together the two tables provide a crude summary of primate population distribution. The data suggest that population density is directly related to elevation, with the greatest density being found above the 150 meter contour, in the rugged and relatively humid highlands that dominate the parishes of St. Peter, St. Andrew and St. Thomas. A lower population density seems to occur in the lower, flatter and drier areas at the northern and southern ends of the island. Likewise, troop sizes appear to be larger in the highlands. Above the 150 meter contour, the mean size of troops reported in the surveys was 12 and the median was 10; below that elevation, mean and median troop sizes were 4.

Reports of monkeys at Government House (Hutson 1977:48), Erdiston College, and the Club Alexandra and Chelston properties, which together occupy a continuous wooded area beside a limestone cliff in Bridgetown, indicate that the monkeys' range includes or adjoins urban neighborhoods. Reports of monkeys in Speightstown, at the Miramar Hotel and in the hamlet of Massiah Street, St. John (Hutson 1977:48) indicate that their occur-

PARISH	SITE	MONKEYS REPORTED(*) OR NUMBER OBSERVED	SITE ELEVATION (METERS)	SOURCE
I. St. Lucy	Archer's Bay	*	10	Denham/Hubl
	Crabhill Plantation	*	10	Denham/Hubl
	Harrison's Point	*	10	Denham/Hubl
II. St. Peter	Speightstown	*	5	Denham/Hubl
	The Whim	*	30	Denham/Hubl
	Sailor's Gulley	3	150	Denham/Hubl
	Mullin's Mill Estate	3	30	Denham/Hubl
	Witchwood Gulley	6	20	Denham/Hubl
	Farley Hill Park	25	210	Denham/Hubl
	Prospect	7+	190	Denham/Hubl
III. St. Andrew	Boscobel	*	120	Denham/Hubl
	Morgan Lewis	*	30	Denham/Hubl
	Cherry Tree Hill	17+	245	Denham/Hubl
	Grenade Hill	3	230	Denham/Hubl
	Turner's Hall Wood	15-20	200	Denham/Hubl
IV. St. James	Miramar Hotel	9	10	DuBois
V. St. Thomas	Welchman Hall Gulley	10	250	Denham/Hubl
	Welchman Hall Gulley	8	250	DuBois
VI. St. Joseph	Joe's River	4-5	120	DuBois
	Ashford Bird Park	*	120	DuBois
	Easy Hall Plantation	*	300	DuBois
	Hackleton's Cliff	*	300	DuBois
VII. St. John	Bath Telegraph Stn.	*	30	DuBois
VIII. St. George	Byde Mill	2-3	100	DuBois
	Drax Hall	4-5	110	DuBois
IX. St. Michael	Flint Hall	3	50	DuBois
	Alexandra Estate	4-5	30	DuBois
	Erdiston College	*	30	Denham/Hubl
	Chelston Estate	*	30	Denham/Hubl
X. Christ Church	Graeme Hall Estate	*	30	Denham/Hubl
	Hopefield Estate	8	60	DuBois
	Yorkshire Estate	4	60	DuBois
	Lowther's Estate	*	60	DuBois
XI. St. Philip	Oughterson Gulley	7	90	DuBois
	Thicket Plantation	2	75	DuBois
	Edgecumbe Plantation	1	60	DuBois
	Halton Plantation	2	90	DuBois
	Kirton Plantation	*	40	DuBois
	Carrington Plantation	*	50	DuBois
	TOTAL	147		

TABLE V. Observations and reports of Barbadian monkeys, 1977-78.

PARISH	NUMBER REPORTED TO TAYLOR	MINIMUM NUMBER OBSERVED BY DuBOIS AND DENHAM/HUBLEY
I. St. Lucy	80	0
II. St. Peter	310	44
III. St. Andrew	220	35
IV. St. James	80	9
V. St. Thomas	175	8
VI. St. Joseph	85	4
VII. St. John	110	0
VIII. St. George	28	6
IX. St. Michael	33	7
X. Christ Church	26	12
XI. St. Philip	27	12
COLUMN TOTAL	1,174	147

TABLE VI. Summary of monkey reports from Taylor (1966) and DuBois and Denham/Hubley (1977-78).

rence in built-up areas may be common, but the immigrants to Nova Scotia who reported monkeys in Speightstown and Black Rock over half a century ago suggest that the monkeys' familiarity with built-up areas may not be a recent development.

Scanty troop composition data make it difficult to generalize about troop structure. About 22% of the monkeys reported in the 1977-78 surveys were adult males, 36% were adult females, 36% were juveniles, and 6% were infants. Counting the Welchman Hall troop only once, six of the seven troops for which good data were collected contained more than one adult male each.

Research Since 1978.
In 1979, Julia Horrocks began a long term study of ecology and social behavior among Barbadian monkeys at Welchman Hall near Taylor's experimental site at Jack-in-the-Box Gully, and in the west coast woodland at Glitter Bay Resort. The project served as the basis for her doctoral dissertation which deals at some length with population size, distribution, and composition (Horrocks 1984).   Also, Horrocks (1983) and Horrocks and Hunte (1983a, 1983b) have published several articles based at least in part on Horrocks' dissertation but dealing primarily with mother-infant relations.

In summary, Horrocks' population figures greatly expand upon but are compatible with those obtained earlier. A detailed survey of 106 farmers yielded an estimated island-wide mean troop size of about 15 and a median size of 14. An early estimate of the total number of monkeys on the island was in the range of 5000 to 7000 (Horrocks 1983) but when a careful census of all the monkeys in St. James Parish revealed a total of 5725 in that parish alone, the estimated total for the island was raised to approximately 12,000 (Horrocks 1984:93-96). Since Horrocks' work focused on the current status of the population rather than on its history, her data say nothing about changes in population size or distribution.

Also in 1979, Jean Baulu received a consultancy contract (CA/1-BS) from the Caribbean Agricultural Research and Development Institute (CARDI) to investigate the extent to which monkeys damage Barbadian crops, and to develop a monkey crop damage control program. The data concerning population size and distribution that he published (CARDI 1982:4-36) were obtained in conjunction with Horrocks' work and agree with her conclusions. Baulu (CARDI 1982:37-51) estimated that monkey damage to sugarcane cost Barbados approximately $4.7 million (Barbadian) in 1980, and that the cost of damage to fruit and vegetable crops that are raised entirely for domestic comsumption was about $3 million (Barbadian). Although these figures are not definitive, they at least sugguest the magnitude of the monkey problem in 1979.

In his discussion of traditional methods of monkey control and their public acceptability (CARDI 1982:51-57), Baulu noted that in a geographically stratified sample of 110 owners or managers of farms where monkeys were pests, only 32 were aware of the government bounty on monkeys, only 28 said that they had killed monkeys during 1979, and only a fifth of those had claimed the government bounty. In an average month the government paid fewer than twenty bounties throughout the island.

Using information that he gained in Barbados as well as prior experience in St. Kitts and elsewhere, Baulu established a monkey control program with financial support from the Caribbean Development Bank and the Barbadian Government. Its objectives were to control monkey damage to crops in a humane fashion, and to make the program financially viable by selling the monkeys for use in polio vaccine production and biomedical research. By June 1981, 363 monkeys had been captured, 325 of them had been exported to Charles River Primates Incorporated in the United States, and other buyers were being sought (CARDI 1982:59-80).

Although the current number of monkeys in Barbados, their basic demographic rates, and the maximum sustainable rate at which they can be trapped without driving the population to extinction remain unknown, the trapping program is continuing. By June 1985, approximately 2,500 monkeys had been exported from Barbados to North America and Europe. In hopes of preventing the extinction of feral monkeys on the island, traps are

used exclusively in areas where the monkeys' numbers and destructiveness are high, and a captive breeding program is being
developed to reduce the pressure on the feral monkeys as their
numbers decline and the export business becomes ever more important to the Barbadian economy. With financial assistance
from the Canadian International Development Agency, Baulu established the Barbados Wildlife Reserve in 1984 as a teaching
and research facility and tourist attraction. It houses a troop
of approximately forty green monkeys (Baulu, pc).

Alternative Interpretations

There are at least three markedly different ways in which
to interpret the Barbadian population data. The "boom-and-
bust" scenario is a refinement and expansion of the traditionally accepted interpretation with which this chapter began.
The "hide-and-seek" scenario suggests that apparent changes in
the size of the green monkey population may be due to more
subtle changes in diet, foraging behavior and visibility. The
"Schomburgk scenario", unlike the other two, assumes that green
monkeys reached Barbados in the 19th century rather than the
17th, and raises questions that differ from those addressed by
the other two.

The Boom-and-Bust Scenario.
The simplest interpretation of the material introduced
above is that green monkeys became established in Barbados
shortly after 1650, and their numbers increased rapidly in the
hospitable, predator-free environment of the woodlands and gullies. By 1680 they became important agricultural pests and a
series of bounty laws was passed, as a result of which the animals were nearly eliminated from the island. Between 1714 and
the 1950's the population remained small, but began to grow
rapidly once again when woods returned to the gullies. By 1975
it approximated the size that it attained three centuries earlier, and the bounty law of 1714 was revived to combat the
pests. This interpretation, which appears in its most complete
form in Taylor's report, is compatible with the existence of
the bounty laws, with comments by Hughes, Chapman, Chester and
Ober, and with part of Schomburgk's report.

Another matter that is compatible with the boom-and-bust
interpretation concerns the general health of the animals.
Hutson (1977:48) suggested that recent suppression of diseases
to which both humans and monkeys are susceptible may have made
an important contribution toward the monkeys' rapid population
growth in recent decades. Such diseases may include poliomye-
litis, tuberculosis, yellow fever and malaria, all of which
have been controlled to some extent in Barbados only since the
middle of the 20th century. Unlike some green monkey populations in Africa that have antibodies to viruses that are related to acquired immune deficiency syndrome (AIDS), an exami-

nation of 125 Barbadian green monkeys showed that none of them possesed those antibodies (Blakeslee, Sowder and Baulu 1985:525). Although some of the monkeys are infected with leptospirosis (J. Baulu and C. Everard, pc), they seem to be in generally good health now; however, evidence concerning changes in their health in recent decades is lacking.

If the boom-and-bust hypothesis is correct, the Barbadian monkeys passed through a narrow population bottleneck that lasted more than 250 years and that could have had significant genetic consequences.

The Hide-and-Seek Scenario.

This alternative assumes that appearances (and disappearances) may be deceiving, and provides an account of events in the history of the Barbadian monkeys that is less direct than the boom-and-bust scenario.

Consider problems of observation. Reporting monkeys is contingent upon observing them, and observing them is contingent upon the appearance of the monkeys, the nature of their habitats, the behavior of both monkeys and observers, and the number of monkeys in the population.

At the present time, it is difficult to observe monkeys in Barbados. The animals live in woods, typically in deep gullies, where visibility is impaired by a combination of dense vegatation and poor lighting, and human mobility is limited by the vegetation and the rugged topography of the gully floors. Even when visibility is good, the size and coloration of the animals makes them inconspicuous. Barbadian monkeys, like those of St. Kitts (McGuire 1974), seem to be very quiet most of the time, and the incessant tradewinds obscure all but the loudest vocalizations. Although the monkeys often venture to the edges of the woods and into canefields, they tend to leave the woods individually and sporadically rather than as entire troops all of whose members can be seen. Furthermore, a troop at the forest-field boundary often "posts a sentinel" who alerts the foragers to possible dangers.

Where shooting is common, monkeys in the open flee from people at distances of 100 meters or more. Baulu's informants reported that they usually could not get closer than 55 meters from the animals, but some said they had been as close as 20 meters without disturbing them (CARDI 1982:19). In the dense vegetation of gullies they promptly disappear if a person tries to sneak up on them, but cautiously peer at one who noisily chops through the undergrowth, stumbles about, or otherwise appears to be inept or inoffensive. Only rarely is it possible to see monkeys from one's automobile while driving beside a canefield or through a gully where a road passes from one limestone terrace to the next. In areas such as the Barbados Wildlife Reserve where the animals are protected, it is easy to approach them within a few meters and they will sometimes initiate friendly direct interactions with visitors.

     With a little practice and persistence, Hubley and I have
seen monkeys frequently in Barbados; on the other hand, one of
our colleagues who grew up in a rural area of Barbados in the
1950's and 60's never saw a single monkey on the island in
twenty years (R. Jordan, pc), and Handler (pc) reports that he
never saw any while he conducted ethnographic fieldwork in the
Scotland District in the 1960's.
     Since it is difficult to observe monkeys in Barbados now,
even when they are considered to be serious agricultural pests,
it is appropriate to be skeptical of early comments concerning
their rarity.   Because of the regrowth of forest in gullies
since 1950, current observation conditions probably are quite
different from those that prevailed between 1665 and 1950, and
it is possible that the earlier, more open habitat would have
enhanced the monkeys' visibility.   On the other hand, it could
have had exactly the opposite effect by making the animals even
more cautious and secretive.   I am not prepared to assert that
this happened, but there is no good reason to simply disregard
its possibility when evaluating reports by casual observers or
scientists who had no special interest in monkeys.   If the
hide-and-seek interpretation is correct, the monkeys were inconspicuous rather than rare for 250 years. This interpretation
receives support from several directions.
     Consider a hypothetical situation in which green monkeys
live in a homogeneous, fine-grained environment. If the availability of resources increases or decreases but remains homogeneous and evenly distributed, changes in the size of the population should be directly related to changes in the resource
supply, but no changes in the spatial distribution of the animals should occur. Conversely, if green monkeys live on an island such as Barbados where the central highlands potentially
provide a much richer habitat than coastal areas, we would pre-
dict that a sharp but uniform decline in the availability of
resources throughout the island would be especially devastating
to animals living in coastal areas.   This proposition suggests
that when the Barbadian monkey population declines, the lowlands might be virtually depopulated before the central highlands lost a significant percentage of their monkeys;   conversely, it suggests that the presence of monkeys in the lowlands might be a useful indicator of a thriving population
throughout the island.
     If the lowland monkeys constitute a sensitive indicator of
conditions throughout the island, data that appear in Figures 5
and 6 offer some support for skepticism concerning reports that
Barbadian monkeys were rare from 1700 to 1950.   The sightings
of monkeys that are plotted in Figure 5 span sixty years and
say nothing about the number of monkeys at each location.   But
they were made by ten independent observers, three of whom explicitly stated that monkeys were rare on the island, and the
distribution of those sightings is similar to that in Figure 6.
It appears that the distribution of the population has remained
stable for the last half-century, and that may mean that its

size has done likewise.

Now consider the bounty. Frequent references to bounty laws notwithstanding, there is no direct evidence that bounty hunting has had a significant impact on the Barbadian monkey population. Although CHAPMAN (1833) hyperbolically refers to both bounty laws and bounty hunting by slaves, three other early reports (HUGHES 1750, SCHOMBURGK 1848, CHESTER 1869) that refer to bounty laws do not mention bounty hunting or sport shooting. Furthermore, OBER (1913:418) specifically warned visiting hunters that monkey shooting was difficult in Barbados. This situation contrasts sharply with frequent early references to monkey hunting in St. Kitts and Nevis (ANDREWS 1931:131-132, COLERIDGE 1825:198, GURNEY 1840:44, GRAINGER 1764:909-910, LABAT 1931:183-185, OBER 1913:340). Finally, Taylor (1966) and Baulu (CARDI 1982) conclude that shooting is an ineffective pest control technique in Barbados. The oft-reported efficacy of the bounty laws seems to be an item of received wisdom that is so plausible that no one bothered to check its accuracy for almost three centuries.

Next consider the availability of trees in gullies and adjacent areas. By linking the recent increase in agricultural damage by monkeys to the recent regeneration of woodlands, Taylor seems to imply that better habitats yield more monkeys, each of whom may do little damage individually but who collectively do more damage than the few monkeys of past years. Although this interpretation sounds plausible and may say something about events in the 20th century, it certainly cannot explain the situation that gave rise to the bounty laws in the 17th century, for those laws were passed a decade or more after most of the woods had been destroyed.

Starkey (1939:75) tried to account for the 17th century monkey problem with an argument that is diametrically opposed to Taylor's (1966) proffered explanation for the 20th century problem. Without citing any evidence to support his position, Starkey suggested that clearing the woods in the 1660's may have caused the monkeys to attack the fields and gardens. Some support for Starkey's position derives from the fact that cane-fields reached their maximum extent on the island between 1680 and 1700, and that the island became extraordinarily dependent on imported provisions during that period (Handler and Lange 1978:89; Starkey 1939:74). With their woodland habitats destroyed, the monkeys might have become more destructive of crops until their numbers diminished somewhat (not due to the bounty, but naturally) to fit the more limited resources, and their destructiveness of food crops might have been especially unacceptable to Barbadians because of the reduced supply of food that was being produced at that time. This hypothesis suggests that the behavior and diet of the monkeys changed when their natural habitats were destroyed, but implies that the 17th century bounty laws were simply wrong in asserting that the number of monkeys on the island had recently increased.

Ironically, Starkey argues that destroying the woods produced the first monkey problem, and Taylor argues that restoring the woods produced the second one. Is there a contradiction here?

It is clear that Taylor's hypothesis cannot apply to the 17th century monkey problem because that problem developed after (or as) the trees were destroyed. Likewise, some evidence suggests that it may not offer a strong explanation of the 20th century monkey problem either. Specifically, while Baulu's position basically agrees with Taylor's, Baulu (CARDI 1982:49) reports that two of the four parishes in which crop damage losses are highest are St. Lucy and St. Phillip, both of which have very low concentrations of gullies (CARDI 1982:16). If crop damage were directly correlated with the availability of gullies and the forests that grow in them, the boom-and-bust scenario (and Taylor's explanation) would be stronger. Such is not the case.

But Starkey's explanation may apply to the 20th century problem as well as to that of the 17th century. Currently Barbados is trying to become self-sufficient in food production, and recent efforts by the Ministry of Agriculture have increased the size and number of vegetable gardens adjacent to hamlets and houses throughout the island. Consequently, people have become increasingly aware of home-grown food, and raiding gardens has become increasingly rewarding for monkeys. Baulu's crop damage report (CARDI 1982), part of which is summarized in Table VII, clearly shows that monkeys damage only a small albeit valuable percentage of the sugarcane, but that they severely damage several kinds of vegetables and fruits.

CROP	% OF CROP DESTROYED ANNUALLY	CROP	% OF CROP DESTROYED ANNUALLY
SUGARCANE	4	FRUITS	
CORN	18	Mango	39
VEGETABLES		Papaya	38
Cucumber	22	Guava	36
Carrot	13	Sugar apple	31
Peas	13	Golden apple	25
Sweet potato	11	Banana	19
Cassava	10	Plum	15
Beans	9	Passion fruit	14
Melon	7	Cherry	12
Cabbage	5	Ackee	11
Tomato	4	Avocado	9
Pepper	4	Breadfruit	3
Yam	3	Citrus fruits	2

TABLE VII. Crop damage by monkeys (from data in CARDI 1982).

Perhaps the risk associated with stealing domestic foods is greater than the potential reward when domestic foods are rare relative to the supply of forest foods, but the potential reward for theft may outweigh the risk when domestic foods are plentiful relative to the supply of forest foods. If so, the monkeys became pests in the 17th century when their supply of forest foods was reduced, under which conditions raiding fields and gardens would have been rewarding to them. Even though the absolute amount of food available in gardens was small then, that amount was large relative to what was in the forest. They became pests in the 20th century when raiding gardens again became rewarding, this time because both relative and absolute amounts of vegetables and fruits available there were large.

In this "hide-and-seek" scenario, changes in the monkeys' diet, foraging behavior and visibility have been misinterpreted as major changes in population size. The hide-and-seek scenario can accommodate a real decline in the 17th century monkey population as a result of destruction of their natural habitat, and it can accommodate a real increase in the 20th century population as a result of reforestation of the gullies. But it shifts the emphasis from monkey numbers to monkey behavior, and suggests that the monkey population of Barbados in the 18th and 19th centuries may have been considerably larger than most reports suggest.

If the hide-and-seek scenario is correct, several important points follow from it. Firstly, the apparent contradiction between Starkey's and Taylor's positions can be reduced; secondly, the likely magnitude of the population bottleneck that the boom-and-bust scenario implies is greatly reduced; and thirdly, the contribution that population growth has made to the current monkey problem must be reevaluated.

The Schomburgk Scenario.

The possibility that Schomburgk correctly identified a New World species in Barbados in the 1840's means that the 17th and early 18th century bounty laws and other available 18th and 19th century reports on Barbadian monkeys may say nothing at all about population changes among green monkeys. Indeed, the bounty laws and Hughes' and Schomburgk's comments may indicate that the New World species that was troublesome in the 17th century finally vanished from the island in the 19th century. If green monkeys were not the villains in the 17th century, and in fact reached Barbados only in the second half of the 19th century, the most pressing questions then relate to the demise of the New World species, the recent arrival of the Old World species, and competition between them if both were present simultaneously. If the Schomburgk scenario is correct, boom-and-bust and hide-and-seek still may be relevant to the history of Barbadian green monkeys, but only to a much more limited extent than would be the case if green monkeys had been there for more than three centuries.

IV. BARBADOS, ST. KITTS AND BEYOND

A Comparison of Barbados and St. Kitts

     Chapter III suggests that intriguing problems wait to be
solved concerning changes in the size of the Barbadian monkey
population that have occurred since 1627.   No such problems
have emerged in my examination of publications that pertain to
Kittitian monkeys.   Kittitian sources convey the overwhelming
impression that monkeys have been extremely numerous and virtually unavoidable since late in the 17th century.   Early works
that pertain to Kittitian monkeys are longer and easier to find
than those that pertain to Barbadian animals (cf. LABAT 1931,
SMITH 1745, GRAINGER 1764, SCHAW 1764 (in ANDREWS 1922), COLE-
RIDGE 1825, GURNEY 1840, and later sources).   Furthermore,
while Kittitian accounts uniformly describe the monkeys as destructive pests, they invariably do so in colorful and humorous
language that is strikingly different from the premature obit-
uaries that characterize the Barbadian sources.
     Although some of the differences in the reports may be due
to differences in the personalities of the writers, other factors may be equally important.   Just as the islands may have
received their monkeys from different sources via different
routes, the situations that the monkeys confronted upon arrival
at the respective islands were not the same.   St. Kitts is a
high, steeply inclined volcanic island; Barbados is an uplift-
ed coral island that is relatively low and flat.   Because of
differences in topography and geology, their plant communities
at the time of European settlement were different, and their
botanical differences were further intensified by the deforestation of Barbados that was not matched in St. Kitts. And since
the end of the 17th century, Barbados has been one of the most
densely populated places in the world, while St. Kitts has remained relatively thinly populated.
     In addition to differences associated with migration his-
tories and natural habitats, Barbadian and Kittitian green monkeys may be more or less similar to each other depending upon
whether boom-and-bust or hide-and-seek more accurately describes population changes in Barbados. In the simplest version
of boom-and-bust, the original and probably small Barbadian
population was growing rapidly in the 17th century when it
crashed under the impact of deforestation and bounty hunting.
When conditions improved in the 1950's rapid growth recurred.
On the other hand, the simplest hide-and-seek scenario suggests
that moderate changes rather than crashes and explosions
characterize the history of the Barbadian population.   Because
of the intense selection pressure implied by boom-and-bust, the
magnitude and rate of genetic change that might have occurred

could have been larger than in the more conservative hide-and-seek scenario. If the commonly accepted boom-and-bust model is correct, differences between Barbadian and Kittitian monkeys may be even larger than differences in migration histories and natural habitats would generate.

The "natural experiment" in which these animals have been presumed to be enmeshed remains unclear. Until better information concerning the history of the monkeys becomes available, it will remain difficult or impossible to determine whether genuine differences between various populations are due to gradualistic divergence from a single parental stock, immigration of two or more discrete founder populations from different parts of Africa, hybridization among two or more subspecies, or rapid change under intense selection pressure on Barbados. Attempts to measure rates and directions of evolutionary change in a natural experiment presuppose accurate knowledge of initial conditions, conditions that characterize the course of the experiment, and elapsed time, none of which is currently available for the West Indian monkeys. And in considering evolutionary problems, it is essential to remember that C. a. sabaeus and its congeners are extraordinarily adaptable sorts of monkeys as demonstrated by their broad geographical and ecological range throughout sub-Saharan Africa. Until someone finds unambiguous evidence that evolutionary change actually has occurred among the West Indian populations, questions about evolution that have permeated much of the research with these animals simply may be premature.

At the very least, we can safely conclude that conspicuous differences in the natural and social histories of Barbados and St. Kitts are incompatible with the notion that the primate populations that live on those islands constitute some kind of unified whole. Because of the many obvious and not-so-obvious differences between the islands, events in the history of one (e.g., passage of the Barbadian bounty laws) must not be integrated with events from another (e.g., Labat's observations of Kittitian monkeys) to yield a synthetic history of West Indian green monkeys as a whole.

Conclusion

Originally I hoped that my inquiry into some facets of the history of the West Indian green monkeys would clarify the course of a potentially important natural experiment. Instead it has raised far more questions than it has answered, but in doing so has focused attention on many problems that have been ignored or misunderstood by biologists for half a century.

By collecting, organizing and publishing the materials that appear in the Appendix, and proposing several alternative interpretations that are superficially compatible with them and the other data at my disposal, I hope I have laid the foundation for more detailed research into the history of the West

Indian cercopithecines. The next and perhaps easiest project would be a thorough examination of unpublished documents that might shed more light on the topics of this monograph. Those sources might radically alter the range of viable hypotheses that I offered in Chapters II and III. Another project would be systematic, comparative field surveys of primate ecology, demography and social behavior among populations of all of the islands, building primarily on Kittitian studies by Sade and Hildrech, Poirier, and McGuire and his associates, and on Barbadian studies by Horrocks and Baulu. Yet another could be field and laboratory studies of genetic similarities and differences among West Indian populations and between West Indian and African populations. Finally there could be comparable studies of the mona monkeys on the island of Grenada. I suspect that there is a useful amount of documentary material concerning them, and it is likely that their history is less complex than that of the green monkeys.

History comes in different sizes and shapes. At one extreme we find micro-historical reports of laboratory research, and at the other we find large-scale long-term inferential histories of Darwin's finches in the Galapagos Islands or honeycreepers in Hawaii. The history that is presented in this monograph lies between those extremes. By being a kind of "human-scale" history of nonhuman animals that is based primarily on human historical documents, it lies somewhat outside the usual scientific tradition. But intriguing, scientifically important events have occurred in the human-scale history of the West Indian cercopithecines, and I suspect that the same is true of many feral animal populations. Since it is a truism that we see different patterns at different scales, this work can serve a useful purpose by beginning to fill the gap between "laboratory time" and "geological time" with regard to the West Indian green monkeys.

REFERENCES

Adanson, M.: A voyage to Senegal, the Isle of Goree, and the River Gambia (J. Nourse, London 1759).

Allen, G.M.: Mammals of the West Indies. Bull. Mus. Comp. Zool. Harvard Coll. 54 No. 6 (1911).

Andrews, E.W., ed.: Journal of a lady of quality 1774-1776 (Yale, New Haven 1921).

Anon.: Larceny of wood. The Barbados advocate. (Bridgetown, Barbados; 5 July 1897).

Anon.: Untitled. The Barbados advocate. (Bridgetown, Barbados; 20 October 1905).

Ashton, E.H.: The influence of geographic isolation on the skull of the green monkey (Cercopithecus aethiops sabaeus). Part V. Proc. Roy. Soc., Ser. B, 151:538 (1960).

Ashton, E.H. and Zucherman, S.: The influence of geographic isolation on the skull of the green monkey (Cercopithecus aethiops sabaeus).
Part I. Proc. Roy. Soc., Ser. B, 137:212-238 (1950);
Part II. Proc. Roy. Soc., Ser. B, 138:204-213 (1951a);
Part III. Proc. Roy. Soc., Ser. B, 138:213-218 (1951b);
Part IV. Proc. Roy. Soc., Ser. B, 138:354-374 (1951c).

Aspinall, A.: A wayfarer in the West Indies (Methuen, London 1927).

Aspinall, A.: Pocket guide to the West Indies (Sifton, Preed and Co., London 1931).

Barbados Department of Archives. Deed to land at Ape's Hill, 18 July 1642. BDA Record Book 3/1:148.

Beard, J.S.: The natural vegetation of the Windward and Leeward Islands. Oxford Forest Memoirs 2 (1949).

Beckwith, M.W.: Jamaica Anansi stories. Mem. Am. Folklore Soc. 17 (1924).

Blakeslee, J.R., Sowder, W.G. and Baulu, J.: Wild African green monkeys of Barbados are HTLV negative. Lancet, 2 March 1985:525.

Booth, A.H.: The distribution of primates in the Gold Coast. W. Afr. Sci. Assn. 2:122-133 (1956a).

Booth, A.H.: The cercopithecidae of the Gold and Ivory Coasts. Ann. and Mag. Natl. Hist. 9:476-480 (1956b).

Buffon, G., comte de.: Histoire naturelle (Paris 1812, 1835).

Burckhardt, J.: The civilization of the Renaissance in Italy (Germany 1860; translated by S.G.C. Middlemore, revised and edited by I. Gordon; Mentor Books, New York 1960).

Burke, J. and C. Caldwell: Hogarth: The complete engravings. (Thames and Hudson, London 1968).

Cameron, T.W.: The species of Subulura Molin. in primates. J.
     Helminthology 8:49-57 (1930).
CARDI (Caribbean Agricultural Research and Development Insti-
     tute):   Monkey crop damage control in Barbados.   Consul-
     tancy contract CA/1-BS. (Caribbean Development Bank 1982).
Chapman, M.J.: Barbadoes and other poems (James Fraser, London
     1833).
Chester, G.J.:   Transatlantic sketches in the West Indies,
     South Atlantic, Canada and the United States (Smith, Elder
     and Co., London 1869).
Clarke, S.:   Account of the four chiefest plantations of the
     English in America (Robert Clavel, London 1670).
Coke, T.: A history of the West Indies, vol. 2 (London 1810).
Coleridge, H.N.: Six months in the West Indies (Murray, London
     1825; reprinted Negro Universities Press, New York 1970).
Colyer, F.: Variations of the teeth of the green monkey in St.
     Kitts.   Proc. Roy. Soc. Med. 41:845-848 (1949).
Curtin, P.:   The Atlantic slave trade: a census (University
     of Wisconsin Press, Madison 1969).
Dandelot, P.:   Note sur classification des Cercopithecus du
     groupe aethiops. Mammalia 23:357-368 (1959).
Daviess, S.: Barbados: a major submarine gravity slide. Geol.
     Soc. Am. Bull. 82:2593-2602 (1971).
Denham, W.W.: History of green monkeys in the West Indies
     Part I. Migration from Africa. J. Barbados Mus. and Hist.
     Soc. 36:211-228 (1981);
     Part II.   Population dynamics of Barbadian monkeys. J.
     Barbados Mus. and Hist. Soc. 36:353-371 (1982).
Duke, W.:   Some memoirs of the first settelment of the island
     of Barbados (Barbados 1741).
Duncan, T.B.: Atlantic islands: Madiera, the Azores, and the
     Cape Verdes in seventeenth century commerce and navigation
     (University of Chicago Press, Chicago 1972).
Dunn, R.S.: Sugar and slaves: rise of the planter class in the
     English West Indies, 1624-1713 (W. Norton, New York 1973).
Edwards, B.: The history, civil and commercial, of the British
     colonies in the West Indies (John Stockdale, London 1807).
Edwards, G.:   Gleanings of natural history (Royal College of
     Physicians, London 1758).
Egerton, F.H.: Egerton Mss. British Museum, London (1780-1820).
Elliot, D.G.: A checklist of mammals of the North American con-
     tinent, the West Indies, and the neighboring seas. Field
     Columbia Museum, Zool. Ser. 6:536-537 (1907).
Ervin, P.M. and McGuire, M.T.:   The St. Kitts vervet (Cerco-
     pithecus aethiops). Carib. Quart. 20:(2):36-52 (1974).
Fowler, H.W.:   Dictionary of modern English usage, 2nd. ed.
     (Oxford University Press, New York 1965).
Fryer, J.: A new account of East India and Persia (London 1698;
     edited by W. Crooke, The Hakluyt Society, London 1909).
Goldsmith, O.:   A history of the earth and animated nature
     (Blackie and Son, Glasgow 1840).
Gooding, E.G.B.:   The plant communities of Barbados (Barbados

          Govt. Printing Office, Bridgetown 1974).
Gooding,   E.G.B.,   Loveless,   A.R. and Proctor, G.R.:   Flora of
     Barbados (Her Majesty's Stationary Office, London 1965).
Grainger,   J.:   Sugar cane.   In A Complete edition of the poets
     of   Great Britain, vol. 10   (Printed for John and   Arthur
     Arch, London 1794).
Gurney, J.: Winter in the West Indies (J. Murray, London 1840).
Hall, E.R.   and   Kelson,   K.R.:   The mammals of North   America
     (Ronald Press, New York 1959).
Hall, R.:   Acts   passed   in the island of   Barbados   1643-1762
     (London 1764).
Handler,   J.S.:   A   guide to source materials for the study of
     Barbados history, 1627-1834. (Southern Illinois University
     Press, Carbondale 1971).
Handler, J.S. and Lange, F.W.:   Plantation slavery in Barbados:
     an   archeological   and historical   investigation   (Harvard
     University Press, Cambridge 1978).
Harlow,   V.T.:   A   history of Barbados   1625-1685.   (Clarendon
     Press,   Oxford 1939;   reprinted Negro Universities Press,
     New York 1969).
Hodgen,   M.T.:   Early anthropology in the sixteenth and seven-
     teenth centuries (University of Pennsylvania Press, Phila-
     delphia 1964).
Hollister,   N.:   The   mona monkey of the island of St.   Kitts.
     Proc. Biol. Soc. Wash. 25:93 (1912).
Horrocks,   J.: Aspects of the behavioral ecology of Cercopithe-
     cus aethiops sabaeus in Barbados.   Ph.   D. thesis (Univer-
     sity of the West Indies, Cave Hill, Barbados, 1982).
Horrocks,   J.   and Hunte, W.:   Maternal rank and offspring rank
     in vervet monkeys.   Anim. Behav. 31:772-782 (1983).
Horrocks,   J. and Hunte, W.:   Rank relations in vervet sisters.
     Am. Naturalist 122(3):417-421 (1983).
Hughes, G.:   The natural history of Barbados (London 1750).
Hutson, L.:   Monkey business.   The Bajan 282:48-49 (May 1977).
Huxley, T.H.:   Man's Place in Nature   (D.   Appleton & Co., New
     York 1898).
Ingersent,   K.A.,   Brathwaite, A.H. and Nurse, J.O.:   Vegetable
     production in Barbados (Barbados Ministry of   Agriculture,
     Bridgetown 1969).
James,   N.P., Mountjoy, E.W. and Omura, S.:   An early Wisconsin
     reef   terrace at Barbados,   West Indies,   and its climatic
     implications.   Geol. Soc. Amer. Bull. 82:2011-2018 (1971).
Janson,   H.W.: Apes and ape lore in the Middle Ages and the Re-
     naissance   (Warburg Institute, University of London 1952).
Jardine, W.:   The naturalists' library: Vol. 1 (mammalia) mon-
     keys (W.H. Lizars and Stirling and Kenny, Edinburgh 1833).
Jekyll,   W.:   Jamaican   song and story (The Folklore   Society,
     London 1906;   reprinted Kraus Reprint Limited,   Liechten-
     stein 1967).
Jukes-Brown, A.J. and Harrison, J.B.:   The geology of Barbados,
     Parts I and II.   Geol. Soc. London Quart. J. 47:197-250
     (1891);   48:170-226 (1892).

Labat, J.-B.:   Nouveau voyages aux isles de l'Amerique (Paris
     1722;   abridged by A. t'Serstevens,   Editions Duchartie,
     Paris 1931).
Ligon,   R.:   A true and exact history of the island of Barbados
     (London 1657;   reprinted Frank Cass Library of West Indian
     Studies No. 11, London 1976).
Linnaeus, C. von: System natura, 12th ed. (Stockholm 1766).
Lowenthal,   D.:   The population of Barbados.   Soc.   and Econ.
     Studies 6:445-501 (1957).
Lucas,   N.:   Lucas Mss.   Barbados Public Library, Bridgetown.
McDermott,   W.C.:   The ape in antiquity (The Johns Hopkins Uni-
     versity Studies in Archaeology No. 27, Baltimore 1938).
McGuire, M.T.:   The St. Kitts vervet (Karger, New York 1974).
Morris,   D.:   British West Indies: geology, climate, vegetation
     and fauna.   In A.J. Horbertson, ed., Oxford survey of the
     British Empire IV: America (Clarendon Press, Oxford 1914).
Moynihan,   M.:   New World primates (Princeton University Press,
     Princeton, N.J. 1976).
Napier,   J.R.   and Napier, P.H.:   A handbook of living primates
     (Academic Press, New York 1967).
Nutting,   C.C.:   Barbados-Antigua Expedition.   University of
     Iowa Stud. in Natl. Hist. 8(3):1-274 (1920).
Ober, F.A.:   Camps in the Caribbees (Lee and Shepard, Boston
     1880).
Ober, F.A.:   A guide to the West Indies (Dodd, Mead and Com-
     pany, New York 1913).
Oliver, P., ed. and trans.:   The voyages made by the Sieur D.B.
     in 1669-1672 (David Nutt, London 1897).
Oxford English Dictionary,   Compact Edition. (Oxford University
     Press, New York 1971)
Parsons,   E.C.:   Barbados folklore.   J. Am. Folklore 38:267-292
     (1925).
Parsons,   E.C.:   Folklore of the Antilles, French and English.
     Mem. Am. Folklore Soc.,   Vol. 26   Pt. 1 (1933),   Pt. 2
     (1936), Pt. 3 (1943).
Pinckard, G.:   Notes on the West Indies (London 1806).
Pocock,   R.I.:   A monographic revision of the monkeys of the
     genus Cercopithecus.   Proc. Zool. Soc. London, Pt.2, 677-
     746, (1907).
Poirier,   F.E.:   The St. Kitts green monkey (Cercopithecus
     aethiops sabaeus).   Folia primatol. 17:20-55 (1972).
Poole,   E.G.   and Barker,   L.H.:   Geology of Barbados.   D.O.S.
     Series 1229, 1:50,000 map with notes (Directorate of Over-
     seas Surveys, London 1983).
Poyer, J.:   The history of Barbados (London 1808).
Public Record Office (PRO):   Calendar of state papers, colonial
     series: America and the West Indies 1574-1738 (London).
Public Record Office (PRO):   Barbados: ships entered and
     cleared, with cargo lists,   1772-1777 (Mss.).   (PRO Doc.
     No. T. 64/48, London).
Ragatz,   L.J.:   A guide for the study of British Caribbean his-
     tory 1763-1834 (U.S. Gov. Print. Office, Washington 1932).

Rawlin, W.: The laws of Barbados (London 1699).

Rickards, C.: The West Indies and Caribbean Yearbook, 48th edition (Caribook Ltd., Toronto 1977).

Sade, D.G. and Hildrech, R.W.: Notes on the green monkey (Cercopithecus aethiops sabaeus) on St. Kitts, West Indies. Carib. J. Sci. 5(1-2):67-81 (1965).

Schomburgk, R.H.: The history of Barbados (Longmans, London 1848; reprinted Frank Cass Library of West Indian Studies No. 19, London 1971).

Schwarz, E.: Die Meerkatzen der Cercopithecus aethiops-gruppe. Z. Säugetierk. 1:28-47 (1936).

Sclater, P.L.: Untitled entry. Proc. Zool. Soc. London 34:79-80 (1866).

Scott, J.: Description of Barbados. Sloane Mss., British Museum 3662, f.62.

Senn, A.: Paleogene of Barbados and its bearing on history and structure of the Antillean-Caribbean region. Amer. Assn. Petrol. Geol. Bull. 5(24):1548-1610 (1940).

Sheridan, R.B.: Sugar and slavery: an economic history of the British West Indies 1623-1775 (Johns Hopkins University Press, Baltimore 1973).

Skeete, C.C.: An historical description of the weather of the island of Barbados, 1901-1960 (Barbados Govt. Printing Office, Bridgetown 1961).

Sloane, H.: Sloane Mss. British Museum, London.

Sloane, H.: A voyage to the islands Madeira, Barbados, Nieves, St. Christopher and Jamaica (Boston 1707).

Smith, J.: The true travels, adventures and observations of Captain John Smith in Europe, Asia, Africa and America, from Anno Domini 1593-1629 (London 1630).

Smith, W.: A natural history of Nevis, and the rest of the Charibee Islands in America (Cambridge 1745).

Starkey, O.P.: The economic geography of Barbados (Columbia University Press, New York 1939).

Struhsaker, T.T.: Phylogenetic implications of some vocalizations of Cercopithecus monkeys. In J. Napier and P. Napier, eds., Old World monkeys, pp. 365-444 (Academic Press, New York 1970).

Tappen, N.: Problems of distribution and adaptation of the African monkeys. Cur. Anthrop. 1:91-120 (1960).

Taylor, K.D.: The rodent and monkey problem in Barbados (Infestation Control Lab., Min. of Agriculture, London 1966).

Trechmann, C.T.: The Scotland Beds of Barbados. Geol. Mag. 62:481-504 (1925).

Trechmann, C.T.: The base and top of the coral rock in Barbados. Geol. Mag. 74:337-358 (1937).

Vernon, K.C. and Carroll, D.M.: Soil and land use surveys no. 18: Barbados (Regional Research Center, University of the West Indies, Trinidad 1965).

Watts, D.: Man's influence on the vegetation of Barbados 1627-1800. Univ. of Hull Occas. Papers in Geog. No 4 (1966).

Williams, G: History of the Liverpool privateers (London 1897).

APPENDIX. HISTORICAL SOURCES

DATE	PLACE	AUTHOR	PUBLISHED
1669	Senegal	Sieur D. B. (Oliver, ed.)	1897
1671	Barbados	John Ried (PRO)	-
1673	Cape Verde	Fryer	1698,1909
1680	Barbados	Legislature (Hall, ed.)	1764
1682	Barbados	Legislature (Hall, ed.)	1764
1684	Barbados	Legislature (Hall, ed.)	1764
1700	St. Kitts	Labat	1722,1931
1714	Barbados	Council (Hall, ed.)	1764
1719	St. Kitts, Nevis	Smith	1745
1750	Barbados	Hughes	1750
1758	Cape Verde	Edwards	1758
1764	St. Kitts	Grainger	1794
1774	St. Kitts	Schaw (Andrews, ed.)	1921
1801	Africa, West Indies	Crow (Williams, ed.)	1897
1807	Barbados	Edwards	1807
1825	Nevis	Coleridge	1825
1833	Cape Verde	Jardine	1833
1833	Barbados	Chapman	1833
1840	St. Kitts	Gurney	1840
1840	Cape Verde	Goldsmith	1840
1848	Barbados	Schomburgk	1848
1866	St. Kitts	Sclater	1866
1869	Barbados	Chester	1869
1880	Grenada	Ober	1880
1880	St. Kitts	Ober (in Hollister)	1912
1903	Barbados	Clark (in Elliot)	1905
1903	Barbados	Clark (in Allen)	1911
1905	St. Kitts, Grenada	Elliot	1905
1907	Barbados, Cape Verde	Pocock	1907
1911	Barbados, St. Kitts, Grenada	Allen	1911
1913	Barbados, St. Kitts	Ober	1913
1914	Barbados, St. Kitts	Morris	1914
1919	Barbados	Nutting	1920
1927	Barbados	Aspinall	1927

Table VIII. Historical sources listed in chrono-
logical order by dates to which sources pertain.

Notes

1.   Table VIII, on the preceeding page, lists sources in
     chronological order by the dates to which they pertain.

2.   In the body of the Appendix, sources appear in alphabeti-
     cal order by author's surname.

3.   For full bibliographic details concerning items appearing
     in the Appendix, see References.

Sources

ALLEN, G.M.: Mammals of the West Indies, 1911:252-253.

     Cercopithecus sabaeus (Linne);   Simia sabaea Linne, Sys.
     Nat., ed. 12, 1776.

     For many years this monkey has lived in a feral state in
Barbados and St. Kitts, where it has been introduced from West
Africa. The time of its introductions is uncertain, but Ligon,
writing of Barbados in 1673 (this refers to the second edition
of Ligon (1657)), did not mention it among the animals of the
island. It probably came sometime during the next seventy-five
years, for Hughes, in 1750, speaks of it in his work on the
natural history of the island.
     Mr. Austin H. Clark, who visited Barbados in 1903, writes
me that owing to the almost complete deforestation of that
place, it is found at only a few points. "In a patch of woodland on the Foster Hall Estate, near Bathsheba, St. Josephs, it
is frequently met with, especially in the early mornings after
a rainy night. At such times the monkeys will often sit on the
larger and more exposed branches of the trees and sun themselves.   I once saw as many as a half dozen on a single large
branch in this wood.   At other times, they are shy and secretive, but if a gun be fired anywhere in the vicinity, it is
almost certain to bring a response in the shape of a bulldog-
like growl from one or more of these animals. Monkeys are also
common in the woods along the upper reaches of Joe's River.
This species is very destructive of fruit grown in the vicinity
of the woods inhabited by it, and will also raid vegetable gardens and sweet potato patches."
     Apparently the monkeys have never been able to increase
very greatly in Barbados.   Hughes (1750) says that they "are
not very numerous ... (they) rather decrease than multiply"
(see HUGHES 1750 for the complete statement).   Schomburgk, in
his history of Barbados written in 1848, states that this monkey was then nearly extinct, but this belief may have been due
to his misapprehension that it was a native species of cebus.
     There seems to be no record of the introduction of this
monkey to St. Kitts, though it is said to have become common.

ANDREWS, E.W., ed.: Journal of a lady of quality (Janet Schaw,
   1774), 1921:131-132.

   She shewed me several fine plantations (in St.   Kitts) ...
(One) is situated rather high, and goes by the name of Monkey
Hill, from which I suppose it more particularly infested by
those gentry, from which indeed no part of the island is entirely free.   As I am no enemy to the Pythagorean system, I do
suppose these lively and troublesome companions (are the suc-
cessors of) the former inhabitants of this island, who you know
were French, and truly the difference is so little between one
Monkey and another, that the transmigration must have been very
easy, and as to the soul, it has undergone no change, but is
French in all respects.   They grin, they laugh, they chatter
and make grimaces.   Their frolicks are mischievous, their
thefts dextrous.   They are subtle enemies and false friends.
When pursued, they fly to the mountain and laugh at their pur-
suers, as they are as little ashamed of a defeat as a French
admiral or general. In short they are the torment of the planters; they destroy whole cane-pieces in a few hours, and come
in troops from the mountain, whose trees afford them shelter.
No method to get the better of them has yet been found out.   I
should think strong English dogs the best; as the English is
the only animal to humble your French monkey and settle his
frolicks.

   (Schaw's reference to the Pythagorean doctrine of metem-
psychosis, or transmigration of souls, bears a striking resemblance to a fragment of The Progress of the Soul, published by
John Donne in 1601, wherein a deathless soul temporarily occu-
pies a monkey's body before coming to rest within the body of
Adam and Eve's daughter (cf. Janson 1952:272-274,284).)

ASPINALL, A.:   A wayfarer in the West Indies, 1927:23.

   A few families of (monkeys) survive ... in the mahogany
grove at Porter's estate, where it is said that they have been
known to take possession of the lawn-tennis court and,
performing the antics of the human players, throw the balls
back and forwards over the net with considerable skill.

CHAPMAN, M.J.:   Barbadoes and other poems, 1833:6, 88.

   The chattering monkey is no longer seen
   To play his pranks among the leafy green;
   Man drove him first from his ancestral wood,
   Then, cruel tyranté thirsted for his blood.
   No more his active form is seen to bound
   From tree to tree, or light upon the ground;
   No more he pelts with nuts his hated foe,

>   Or scolds at him that stands and laughs below -
>   The cunning miniature of man is gone,
>   Slain in the empire which once was his own! (pg.6)

One cannot help regretting the extermination of this harmless and amusing family. The negroes had the most extravagant notions on the subject of their sagacity. It was a saying of their's, "Jacko know for talk well enough; but he too cute for talk; 'spose he talk, massa buckra make he work." Their admiration, however, could not resist the temptation of slaying Jacko, on whose head the price of a dollar was set. (pg.88)

(The paragraph from page 88 is a footnote to the verse on page 6. Chapman's comment concerning speech by monkeys, which he attributes to "the negroes", is a version of a common European folktale that depicts monkeys as "mute philosophers". In addition to its popularity in oral tradition, it appears in the published works of people such as Alexander Pope, Lord Monboddo and Charlotte Bronte. See Janson (1952:337-338, 351) for other variants and numerous references.)

CHESTER, G.J.: Transatlantic sketches, 1869:39, 49.

The gullies form the most striking and peculiar feature in Barbadian scenery. ... Some of the gullies also afford a refuge for the raccoons and monkeys, which advancing cultivation and rewards for their extirpation now render extremely rare. (pg.39)

Five land animals only are indigenous to Barbados: of these the monkey and the raccoon are rarely met with and are confined to the undercliff of the Scotland District, and to a few gullies. (pg.49)

COLERIDGE, H.N.: Six months in the West Indies, 1825:198.

In Nevis a man is always placed as sentinel in a pinery, for otherwise those dogs, the monkeys, who are very good livers and know a ripe pine (i.e., pineapple) to a day, are sure to take an evening walk from the mountain, and will, I am told, fairly pick, pack, and carry away all the eatable fruit in a garden at one visit.   Certainly Jacco is a rogue, a villain, a thief, yet the fellow's cleverness is so great, his malice so keen, his impudence so intense, that it exceeds the hardness of my heart not to like him. You may offer your fine green Seville oranges to him by handsfull;   deuce o' bit of the rind of ten thousand of them will Jacco touch; no! no! massa - dem monkies savey what bitter as well as buckra!

(Although the comment about monkeys' refusing to eat fruit
with a bitter rind sounds like a naturalistic observation and
indeed   may   be,   it   also bears a striking resemblance to a
ancient   fable that was popular in several   European   languages
throughout the later Middle Ages:

> This   tale,   which seems to have been invented soon
> after 1200,   informs us that apes (i.e.,   monkeys)
> are very fond of nuts;   yet when (one) finds a nut
> with a bitter rind,   he will throw it away, because
> he is afraid to face the unpleasant task of biting
> through   it,   so that he never discovers the   sweet
> kernel inside.   The rind,   needless to add, stands
> for   the   trials and tribulations which   the   pious
> Christian   must shoulder during his life on   earth,
> while   the kernel represents the   heavenly   reward.
> Odo of Cheriton,   our earliest known source for the
> story,   links it with a quotation from St. Gregory:
> "The   stupid   one   prefers   eternal   punishment   to
> temporary adversity"   (Janson 1952:148).)

EDWARDS, G.:   Gleanings of natural History, 1758:10-11.

The St. Jago Monkey. The monkey is often called the green
monkey, and is known to us by that name. Our seamen generally
call them St. Jago Monkies, they being brought from St. Jago,
one of the Cape de Verde islands, lying off the Cape de Verde,
in the Atlantic ocean, from fifteen to eighteen degrees of
north latitude. (pg.10)

I once had an opportunity of seeing, in the house of the
late Duke of Richmond, at Whitehall, an old she monkey, who had
been brought to England with young. ... This sort of Monkey
being pretty commonly brought into most of the maritime trading
ports of Europe, it has probably been described by some former
naturalist; though I can find no figure, that will answer better to this, than it will to several other sorts of monkey.
(pg.11)

(This text accompanies the original color plate that is
reproduced in black-and-white on the cover of this monograph.)

ELLIOT, D.G.:   A checklist of mammals, 1905:536-537.

1308. Cercopithecus callitricus Is. Geoffroy. Green Monkey
Type locality. Senegambia; West Africa.
Geogr. Distr. West Africa, Senegambia to River Niger. In-
troduced into the islands of St. Kitts and Barbados, West
Indies.

Not having been able to obtain any reliable information
regarding the species in the West Indies, it was omitted from
the work on the mammals of Middle America and the West Indies.
I have lately been informed by Mr. A. H. Clark of Boston that
it is common in certain districts of Barbados and also in St.
Kitts, and is, therefore, included in this list. I have heard
that another species of monkey from the Old World has been introduced into the Island of Grenada, but although I have tried
to obtain some information concerning it, and written to persons on the Island, yet I am unable to give its name or learn
anything about it.

FRYER, J.:   A new account of East India and Persia, 1909:40.

(The following report pertains to St. Jago Island in the
Cape Verdes, 9 February 1673.)

Here met us whole Troops of the Natives with their several
Wares, some offering us Cocoes, others Oranges and Limes; some
brought Jackanaps's, such green Ones as are commonly seen in
England to be sold; and all at the price of a cleanly Rag, or
a Bunch of Ribbons.

GOLDSMITH, O.:   A history of the earth, 1840:507.

The sixth is the Callitrix, or Green Monkey of St. Jago,
distinguished by its beautiful green color on the back, its
white breast and belly, and its black face.
As this monkey is found in Cape de Verde islands and the
neighboring parts of Africa, it is one of the species most
frequently imported into Europe.

GRAINGER, J.:   Sugar cane, 1794:909-910.

(The following lines appear in Book II of a didactic poem
that Grainger wrote while living in St. Kitts. He heavily an-
notated the poem, and the footnote accompanies the original
text. The content closely resembles some of Janet Schaw's
remarks about the French (see ANDREWS above).)

The monkey nation preys: from rocky heights,
In silent parties, they descend by night,
And posting watchful sentinels, to warn
When hostile steps approach; with gambols they
Pour O'er the cane grove. Luckless he to whom
That land pertains! in evil hour, perhaps,
And thoughtless of tomorrow, on a die
He hazards millions; or, perhaps, reclines
On luxury's soft lap, the pest of wealth;

And, inconsiderate, deems his Indian crops
Will amply her insatiate wants supply
From these insidious droll (peculiar pest
Of Liamuiga's hills) would though defend
Thy waving wealth; in traps put not thy trust,
However baited: Treble every watch,
And well with arms provide them; faithful dogs,
Of nose sagacious, on their footsteps wait.
With these attack the predatory bands;
Quickly the unequal conflict they decline,
And chattering fling their ill-got spoils away.
So when, of late, innumerous Galiic hosts
Fierce, wanton, cruel, did by stealth invade
The peacable American's domains,
While desolation mark'd their faithless rout;
No sooner Albion's martial sons advanc'd,
Than the gay dastards to the forests fled
And left their spoils and tomahawks behind.

Ver. 46. The monkies which are now so numerous in the
mountainous parts of St. Christopher, were brought thither by
the French when they possessed half the island. This circum-
stance we learn from Pere Labat, who further tells us, that
they are a most delicate food. The English negroes are very
fond of them, but the white inhabitants do not eat them. They
do a great deal of mischief in St. Kitts, destroying many thousand pounds Sterling's worth of canes every year.

GURNEY, J.: A winter in the West Indies, 1840:44.

The highlands of St. Christopher's ... are clothed with a
forest of hardwood ... (that) abounds with monkeys, mischievous
enough among the canes, but always too cunning to be caught or
shot. They regularly employ a sentinel in advance, who sets up
a terrible screeching as soon as danger approaches.

HALL, R.: The laws of Barbados, 1764:106, 111-112, 215.

(The first act that is reproduced below does not appear in
Hall (1764). Rather, it is of uncertain provenance, is dated
17 March 1679, and was provided to me by Mr. Warren Alleyne of
Bridgetown, Barbados. It may be an incorrectly dated copy of
the act that was passed on 17 March 1680. All four of the
"monkey laws" are reproduced here in full, primarily to
demonstrate how little they actually say about Barbadian
monkeys in the 17th and 18th centuries.)

An Act for Destroying Wilde Monkeys and Raccoons.   Provenance unknown; dated 17 March 1679 (1680?).

Whereas the monkeys that have been wilde in the woods have
of late years increased to such great numbers that many of the
inhabitants are at present much damnified by them and time all
may, therefore to prevent the future loss or prejudice; be it
enacted by His Excellency, Sir Jonathan Atkins, Kt., Captain
General and Chiefe Governor of this and other the Charibee Islands the councell and the representatives of this island, that
what person soever shall after the publication of this act,
take or kill any wilde monkey or raccoone to either of the
churchwardens belonging to the parish wherein the monkey or
raccoone was killed, shall receive immediately five shillings
from the said churchwarden for each wild monkeys or raccoones
head soe produced, and the respective churchwardens are hereby
required to pay the same, and hereby allowed to charge the same
to the parish accompt, and the respective vestryes of this island are hereby impowered to make a leavy in their respective
parishes to defray the charge thereof.

An Act for Destroying Wild Monkeys and Raccoons.   Hall,
1764:106; passed 12 July 1682.

Whereas the wild monkeys and the raccoons in several parts
of this island have lately much increased to the great damage
of many inhabitants, and if due care be not taken to prevent
them, they may in a short time become so numerous that the
whole island may be much infested by such vile creatures;   be
it therefore enacted by His Excellency, Sir Richard Dutton,
Knight, Captain General and Chief Governor of this and other
the Charibbee-Islands, and by authority of the same, that what
person soever shall, after the publication of this act, take or
kill any wild monkey or raccoon, and carry the head of the said
monkey or raccoon to any of the churchwardens belonging to the
parish wherein the monkey or raccoon was killed, shall immediately receive five shillings from the said churchwarden, for
such wild monkey's or raccoon's head so produced; and the respective churchwardens are required to pay the same; and hereby allowed to charge the same to their parish account. And the
respective vestries of this island are hereby empowered to make
a levy in their respective parishes to defray the charge thereof. This act to continue two years and no longer.

An Act to Revive and Continue an Act Entitled "An Act for
Destroying Wild Monkeys and Raccoons".   Hall 1764:111-112;
passed 6 November 1684.

Be it enacted by His Excellency, Sir Richard Dutton,
Kinght, Captain General and Governor in Chief of this and other

the Charibbee Islands, the Honorable Council and the General
Assembly of this island, that an act bearing the date the thir-
tieth day of June one thousand six hundred and eighty two,
entitled "An Act for Destroying Wild Monkeys and Raccoons", and
all and every the authorities, clauses, branches, penalties and
provisoes in the said act contained and enacted, be and hereby
declared to be revived and continued, and have their full
force, strength and virtue according to the true intent and
meaning thereof, notwithstanding any limitation therein mentioned.

> An Additional Act to an Act Entitled "An Act for Destroy-
> ing Wild Monkeys and Raccoons".   Hall 1764:215; passed 4
> March 1714.

Whereas in an act entitled "An Act for Destroying Wild
Monkeys and Raccoons" in this island, there having been no pen-
alty provided for, against the churchwardens for refusing payment of the sums therein mentioned;   to any such persons as
should kill and take wild monkeys and raccoons, and several
persons having at time carried wild monkeys and raccoons to
several of the churchwardens of this island, pursuant to the
said law, who have refused payment of the same;   be it therefore enacted by the Honorable William Sharpe, Esq., President
of His Majesty's Council, and Commander and Chief of this and
all the Charibbee-Islands windward of Guadeloupe etc., the Honorable the members of His Majesty's Council and the General Assembly of this island, and by the authority of the same, that
upon any churchwardens refusing payment of the sums mentioned
in the said act, for killing and destroying wild monkeys and
raccoons he shall forfeit and pay the sum of ten shillings to
the informer;   to be recovered before the next justice of the
peace, as in case of servants wages.

HOLLISTER, N.:   The mona monkey of ... St. Kitts, 1912:93.

There is, apparently, no published record of the Mona monkey on the island of St. Kitts, West Indies. In Dr. Glover M.
Allen's recently published Mammals of the West Indies, the
Green Guenon is recorded from this island, as well as from Barbados, and the Mona is recorded from Grenada. In the United
States National Museum collection, in addition to specimens of
the Green Guenon, is a skin of the Mona (Lasiopyga mona) collected on St. Kitts in 1880 by Mr. Fred S. Ober.

HUGHES, G.:   The natural history of Barbados, 1750:66.

Monkeys.   These are not very numerous in this island.
They chiefly reside in inaccessible gullies;   especially where

there are many fruit trees.

The greatest mischief that they do to the neighboring planters is digging out of the earth their yams and potatoes, and sometimes breaking and carrying off a great many ripe sugarcanes.

As a law of this island provides a premium for destroying these, as well as raccoons, they yearly rather decrease than multiply.

JARDINE, W.: Monkeys, 1833:140-142.

The green Monkey.
   St. Jago monkey, Edward's Gleanings, plate 215.
   Green monkey, Pennant's Quadrapeds, 203.
   Guenon callitriche, Demarist's Mammalogie, 61.
   Cercocebus sabaeus, Geof. Annals du Museum, 19:99.
   La callitriche, F. Cuvier Hist. Nat. des Mammifers.

The green monkey is one of the most abundant of (the cercopithecines), and is perhaps oftener seen in a captive state than any of the others. It is a native of the Cape Verde islands and the continent of Africa.

This animal, or at least one under the title of "Green Monkey", has been mentioned by many travellers, who give accounts of the vast troops which accumulate together. In Adanson's Voyage to Senegal, it is thus introduced;

   The place was very woody, and full of green monkeys, which
   I did not perceive but by their breaking the boughs on the
   tops of the trees, from which they tumbled down upon me;
   for in other respects they were so silent and nimble in
   their tricks, that it would have been difficult to hear
   them.

(The quotation from Adanson (1759) appears in Buffon (1767), but Buffon does not cite Adanson as his source. Elsewhere, Jardine (1833:211) reports that green monkeys inhabit the island of Mauritius in the Indian Ocean. Finally, Jardine discusses cebus monkeys at some length on pp.219-222.)

LABAT, J.-B.: Nouveau voyage, vol 2, 1931:42-43, 183-185.

The slave trade is not the only commercial activity carried on along the African coasts. There is also considerable trade in gold, elephant tusks, called morfil, wax, leather, gum, and malaquetta pepper. Parrots, monkeys, fabric or grass skirts and other things also are exported.

Concerning monkeys, I was once told of an incident by an officer of one of these Companies in which his father had been involved while he was main clerk at one of their trading set-

tlements.   It is too pleasant to forget, although I cannot
vouch for its reliability because it was told to me by a person
of whom I know very little.

This clerk, having asked for leave to journey to France to
handle private matters, received an order from one of the
general directors asking him to bring four or five monkeys;
having spelled out the quantity rather than writing it in
numerals, he had actually requested four or five hundred monkeys.   The poor clerk did not know what to think of such a request,   nor did he know into which country these animals would
be introduced.   He went to a great deal of trouble to amass so
many monkeys and to have cages and huts prepared in which to
keep them on the ship.   In spite of his efforts he could not
find the specified amount;   he had to settle for about three
hundred and thirty which he shipped and which,   with the excep-
tion of those that fell into the sea,   arrived safely at La Ro-
chelle.   The clerk immediately went to see the director who had
written to him;   when asked whether he had brought the requested monkeys,   the poor, trembling clerk answered that he had not
been able to fill the order completely and that some had fallen
into the sea during the crossing, so that only about three hundred and ten remained. One can imagine the director's surprise;
he became very angry at the clerk and told him that he had asked for only four or five monkeys,   that if he had brought more
it would be at his own expense,   and furthermore, that he would
be made to pay for the damages caused to the Company by such a
cargo.   Realizing where this could lead,   the clerk handed the
director's letter over to the court registrar for safekeeping
and had him notarize a collated copy.   Upon seeing, in his own
handwriting,   the actual request for four or five hundred monkeys, the director was then obliged to take care of the beautiful merchandise which served as magnificent presents for his
colleagues and friends (pg.42-43).

That evening (on the island of St. Christopher) we were
entertained in a way that I had not expected,   which was to go
monkey hunting.   While the English had control of the French
land,   most of which remained fallow, the monkeys which escaped
from the houses of the French during the war,   multiplied so
quickly that they were seen in large troops when possession of
the island was retaken.   They would even come into the houses
to steal,   and when cane,   potatoes,   and other things were
planted,   it was necessary to be on guard day and night to
prevent the animals from taking everything that had been put
into the soil.

At Mr. Lambert's,   cane was planted in fields quite close
to the round mountain,   one of the hideouts of the animals.   We
went to hide in this area about an hour before sunset.   After
less than an hour we were happy to see a big monkey come out of
the bushes.   After carefully looking around him, he climbed into a tree from which he inspected the surrounding area.   Finally, he uttered a scream,   at once answered by over a hundred

different voices, and immediately afterwords we saw the arrival
of a large troop of monkeys of different sizes which jumped into the canefield and started to tear cane out and carry it
away. Some of them took four or five pieces which they put on
one shoulder and left, jumping on their hind legs; others took
one in their mouths and left walking on all four legs. We shot
after having watched their antics long enough. We killed four
of them, one of which was a female with her baby still clinging
to her back. It held her in the same way our little negroes
hold their mothers. We took the little one with us and raised
it, and it became the prettiest animal one could have desired.
    It was on this occasion that I ate monkey for the first
time. It is true that at first I felt some reluctance when I
saw four heads, which resembled little children's heads, float-
ing on the soup; but as soon as I tasted it, I easily changed
my mind and continued to eat with pleasure, for it is a tender,
delicate white meat, very juicy, and equally good with any kind
of sauce.
    Concerning this little monkey, an incident happened to
Father Cabasson that is worth mentioning here. Having been
raised by Father Cabasson, the little animal grew so affection-
ate that it would never leave him; therefore, it became necessary to lock him inside carefully every time the Father went to
church, since he had no leash on which to keep him. He escaped
once, hid above the pulpit, and came out only after his master
had begun the sermon. He sat down on the edge and, looking at
the priest's gestures, immediately imitated them with faces and
poses that made everybody laugh. Father Cabasson, not under-
standing the basis for the irreverence, at first ignored the
laughter, but as it increased he became very angry and began
condemning quite strongly the lack of respect for God's words.
His gestures, more exaggerated than before, only increased his
monkey's faces and poses and the congregation's laughter.
Finally somebody suggested that the priest look above his head
to see what was happening. No sooner had he seen his monkey's
antics than he began laughing himself and since there was no
way to get ahold of the animal he preferred to give up his sermon, he being unable to continue and his audience unable to
listen (pg.183-185).

MORRIS, D.: British West Indies, 1914:376.

    The Barbados and St. Kitts monkey (Ceropithecus callitri-
cus), now established in these islands, was introduced from
Sierra Leone in slavery times.

NUTTING, C.C.: Barbados-Antigua expedition, 1920:66.

    Mammals. We paid practically no attention to land verte-
brates at Barbados, as they are few and well known. There is a

species of monkey in a small patch of woodland in the northern part of the island, but we did not secure specimens for identification.

OBER, F.S.: Guide to the West Indies, 1913:337-340, 418-419.

Concerning St. Kitts
The only inhabitants at present of the fortress on Brimstone Hill are the wild monkeys, with which the forests above abound, as they occasionally stray to the lowlands. Good monkey-hunting, by the way, may be had in the great forests that surround Mount Misery; but the animals are so exceedingly shy that few of them are ever shot. Almost any sugar planter living on the slopes of the mountain can put one in the way of gratifying a desire for slaying a simian, or, at least, can direct him to the animals' haunts, which are in the high woods generally, with frequent forays into the plantations (pg.337-338).

All over the island range the ubiquitous wild monkeys, even in the region about the Salt Pond (reached by boat from Basse Terre), where there is little cover to shield them from the hunter. The big forests, however, are preferred by the monkeys as their haunts; and they are said to make use of a subterranean passage beneath the sea channel between St. Kitts and Nevis, to range from one island to the other (pg. 340).

Concerning Turner's Hall Wood, Barbados
There is little shooting in the island, except of plover and such birds in the winter season, and these woods hold the only wild animals, containing as they do specimens of raccoons and monkeys. They are difficult to obtain, however, for at one time in the past a bounty was placed upon their heads, and they came near being extirpated (pg. 418-419).

OLIVER, P., ed.: The voyage made by the Sieur D. B., 1897:9.

There are settlements of the French at Senegal belonging to the gentlemen of the Company of the West Indies, who trade at this place, whence they obtain Golddust, Ambergris, Musk, Ivory, Skins, Parrots, & Monkeys.

POCOCK, R.I.: A monographic revision of the ... genus Cercopithecus. 1909:726-727, 731-733.

Cercopithecus sabaeus Linn.
Loc. Senegambia, Sierra Leone, and Northern Liberia. Also introduced into some of the Cape Verde and West Indian islands, and still abundant at least in Barbados. This is one of

the commonest monkeys in European menageries (pg.726-727).
    Cercopithecus tantalus Ogilby.
    <u>Loc</u>. Nigeria (Lakaja, Dahomey, Upper Benue River) up to
Lake Chad.
    A large number of specimens of this species have been ex-
hibited in the Society's Gardens. ... Although <u>C. tantalus</u>
has never, I believe, been previously identified with certainty, I do not think there is any reason to doubt the correctness
of my determination. ... I find it impossible to believe that
so common a monkey in menageries has escaped naming down to the
present time. For many years there has been a stuffed example
in the British Museum labelled <u>C. callitrichus</u>; and it was
probably this specimen that caused Dr. Forbes to describe <u>C.
callitricus</u> (i.e., <u>C. a. sabaeus</u>) as having a white brow-band.
<u>C. tantalus</u>, as here recognized, will be found to intergrade.
Up to the present time, however, I have not seen any specimen
that could not with certainty be assigned either to one or the
other of these forms (pg.731-733).

PUBLIC RECORD OFFICE (PRO) Calendar of state papers, colonial
    series, 1669-1674:243.

    August 2, 1671: Barbadoes: John Reid to (Sec. Lord Ar-
lington?). Takes this sure convenience by his honor's old
acquaintance, Capt. Barrett, to let him know he has another old
servant, acquaintance and beadsman alive here. This island
affords nothing worthy of his Lordship's acceptance, but has
delivered Capt. Barrett a monkey to be presented to her Lady-
ship, being confident it will please her for it is the finest
he ever saw.

SCHOMBURGK, R.: The history of Barbados, 1848:682-683.

    If we except domestic animals, Barbados possesses five
genera of terrestrial animals, comprising only a few more species in number. The most interesting is the Barbados Monkey,
now nearly extinct, although formerly so frequent that the
Legislature set a price upon its head. I have much to regret,
on account of natural history, that my endeavors to procure a
specimen for purposes of determining the species have entirely
failed. From the outer appearance of a living specimen, I consider it to be <u>Cepus capucinus</u>, Geoff., the Sai or Weeper, or a
very closely-allied species. It is not likely that it was introduced, as the first settlers found it in large numbers on
their arrival. The Raccoon, <u>Procyon lotor</u>, Cuv., is now equally scarce, although formerly so abundant that they were included in the legislative enactment for extirpation. If we add
to this two animals, perhaps an indigenous mouse and two species of bats, we come, as far as my knowledge extends, to the
end of our enumeration of indigenous mammalia.

SCLATER, P.L.: Untitled entry, 1866:79-80.

     Mr. Slater called the attention of the meeting to three
Monkeys recently received from the Island of St. Kitts, West
Indies. ... The animals were undoubtedly referable to the common Green Monkey (Cercopithecus callitricus, Geoff.) of Western
Africa, and must have been introduced years ago, as they were
stated to be now very abundant in the woods of St. Kitts, and
to cause great damage to the sugar-plantations.

SMITH, W.: A natural history of Nevis, 1745:35.

     (I was unable to obtain a copy of this book.   The
following is quoted from Sade and Hildrech 1965:68-69.)

     In May 1719 the Reverend William Smith of Nevis visited
St. Kitts, and in May 1745 he published a collection of letters
to a friend in England. He describes a climb to the great
crater below Mt. Misery, and mentions in parentheses that the
thick woods

          "swarm with wild Monkies who venture down in the
          dark night to steal potatoes and other provisions
          with so much cunning or craft as to give rise to
          several strange incredible Stories about them ..."
          Smith (1745:35).

Unfortunately he does not relate any of the ... stories, nor
are monkeys mentioned again in his book. ... (A)lthough he describes the fauna of Nevis he makes no mention of monkeys.

WILLIAMS, G.: History of the Liverpool privateers, 1897:384,
     647-648, 651, 654.

     On the 5th of March, 1801, the Bolton ... on her passage
from Demarara (Guyana) for Liverpool, engaged for an hour ... a
large French privateer. ... In addition to a valuable cargo of
sugar, coffee, cotton, elephant teeth, etc., which was plundered by the privateer, the Bolton had a very fine tiger on board,
and a large collection of birds, monkeys, etc. (pg.384).

     Captain Hugh Crow, (the commander of) the last slave ship
that cleared out of the port of Liverpool (in 1807) ... tells a
humerous story of a monkey who wanted to take command of the
Mary.... He was uncommonly expert in imitating anything he saw
done, particularly if it were mischievous.... (Captain Crow
and the monkey came into conflict concerning the use of the
speaking trumpet while Crow was trying to prepare the ship for
a storm.   A scuffle ensued.) "Before I had time to look about
me, the fellow sprang at my neck, and after chattering and mak-

ing faces of great consequence, he bit me several times". (The
monkey was chained, but some days later escaped to the Captain's cabin and completely ransacked it.) "Owing to these and
similar pranks, I determined to part with him, and a few days
after we arrived in Kingston (Jamaica) I had him advertised in
the newspapers". (He dressed the monkey in clothes and sold
him.) "Next morning the wags in the town reported ... that he
had run off with two half firkins of butter from a provision
store, and would certainly be tried and banished the colony for
so grave an offence" (pg.647-648).

After leaving Bonny with a load of slaves in 1807, Captain
Crow put into the Portuguese island of St. Thomas. ... (Later)
the ship resumed her voyage with additional passengers in the
shape of several monkeys presented by the governor to Captain
Crow (pg.651).

"On this passage" (Crow says), "we had several monkeys on
board. They were of different sizes and species, and amongst
them was a beautiful little creature, the body of which was
about ten inches or a foot in length, and about the circum-
frence of a common drinking glass. It was of a glossy black,
excepting its nose and the end of its tail, which were white as
snow. This interesting little animal, which, when I received
it from the Governor of the island of St. Thomas, diverted me
by its innocent gambols, became afflicted by the malady which
yet, unfortunately, prevailed in the ship". (In the remainder
of this lengthy passage, Crow describes in some detail the care
bestowed upon the dying monkey by the other healthy monkeys.
The passage implies, but does not state, that all of the animals except the little black-and-white one arrived safely in
the New World.) (pg.654, fn.)

Subject Index

Note: This index is highly selective, especially with regard to
the names of people, places and species.